AF294865

Springer Series in
Computational
Mathematics

# 3

N. Z. Shor

# Minimization Methods for Non-Differentiable Functions

Translated from the Russian
by K. C. Kiwiel and A. Ruszczyński

Springer-Verlag
Berlin Heidelberg New York Tokyo

Naum Zuselevich Shor
Institute of Cybernetics of the Academy of Sciences
of the Ukrainian SSR, 142/144 40-letiya Oktyabrya Avenue
25227, Kiev-207, USSR

Krzysztof C. Kiwiel
Systems Research Institute, Polish Academy of Sciences,
ul. Newelska 6, 01-447 Warsaw, Poland

Andrzej Ruszczyński
Institute of Automatic Control, Technical University of Warsaw,
ul. Nowowiejska 15/19, 00-665 Warsaw, Poland

Title of the Russian original edition:
Metody minimizatsii nedifferentsiruemykh funktsij
i ikh prilozheniya
Published by Naukova Dumka, Kiev 1979

AMS Subject Classifications: 26B25, 49D07, 49D27, 49D37, 52A40,
65H10, 65K05, 90C06, 90C30

ISBN-13: 978-3-642-82120-2        e-ISBN-13: 978-3-642-82118-9
DOI: 10.1007/978-3-642-82118-9

Library of Congress Cataloging in Publication Data.
Shor, Naum Zuselevich. Minimization methods for non-differentiable functions.
(Springer series in computational mathematics ; 3) Translation of: Metody minimi-
zatsii nedifferentsiruemykh funktsii i ikh prilozheniia. Bibliography: p. Includes in-
dex. 1. Mathematical optimization. 2. Nondifferentiable functions. I. Title. II. Series.
QA402.5.S5413    1985    519    84-23594



Typesetting: Graphischer Betrieb Konrad Triltsch, Würzburg
Bookbinding: B. Helm, Berlin
2141/3020-543210

# Preface

In recent years much attention has been given to the development of automatic systems of planning, design and control in various branches of the national economy. Quality of decisions is an issue which has come to the forefront, increasing the significance of optimization algorithms in mathematical software packages for automatic systems of various levels and purposes. Methods for minimizing functions with discontinuous gradients are gaining in importance and the experts in the computational methods of mathematical programming tend to agree that progress in the development of algorithms for minimizing nonsmooth functions is the key to the construction of efficient techniques for solving large scale problems.

This monograph summarizes to a certain extent fifteen years of the author's work on developing generalized gradient methods for nonsmooth minimization. This work started in the department of economic cybernetics of the Institute of Cybernetics of the Ukrainian Academy of Sciences under the supervision of V.S. Mikhalevich, a member of the Ukrainian Academy of Sciences, in connection with the need for solutions to important, practical problems of optimal planning and design.

In Chap. 1 we describe basic classes of nonsmooth functions that are differentiable almost everywhere, and analyze various ways of defining generalized gradient sets.

In Chap. 2 we study in detail various versions of the subgradient method, show their relation to the methods of Fejer-type approximations and briefly present the fundamentals of $\varepsilon$-subgradient methods.

In Chap. 3 we describe gradient-type algorithms with space dilation in the direction of the gradient (the SDG algorithms) and in the direction of the difference of two successive gradients (the $r$-algorithms). Great attention is paid to establishing convergence and speed of convergence of these algorithms. Extreme variants of the methods are also considered. We separately analyze the applicability of the SDG algorithms to the solution of systems of nonlinear equations. The relations between cutting plane methods and methods with space dilation are also discussed.

Chapter 4 is devoted to the use of generalized gradient methods for implementing decomposition schemes in numerous real-life problems of operative and long-term planning, for minimax problems, for problems of interpreting gravimetric observations, etc.

The author expresses his sincere gratitude to V.S. Mihkalevich, L.A. Galustova, G.I. Gorbach, V.I. Gershovich, N.G. Zhurbenko, T.V. Marchuk, A.I. Momot, V.A. Trubin, L.P. Shabashova, and G.N. Yun, who contributed significantly to the elaboration and practical applications of the methods of nonsmooth optimization reflected in this monograph, and to V.N. Bogdanyuk, G.A. Verbe, T.G. Thrushko and T.I. Yatsenko for their assistance in preparing the manuscript.

# Contents

# Introduction

Nonsmooth functions, i.e. functions with discontinuous gradients, arise frequently in the theory and applications of mathematical programming. As far as the theory is concerned, it is now customary to consider basic facts of convex analysis for the whole class of convex functions, including nonsmooth ones; the use of subdifferentials has become as common as the use of gradients, especially after the fundamental work of R. T. Rockafellar [70], B. N. Pshenichny [68] and others.

There exist several principal sources of nonsmooth optimization problems. Firstly, one may point out large scale mathematical programming problems with block structure and a relatively small number of interactions between the blocks. An application of decomposition methods to such problems yields problems of minimizing (maximizing) functions that usually depend in a nonsmooth way on the linking variables or the Lagrange multipliers (shadow prices) corresponding to the interactions.

Secondly, we may mention minimization problems involving max functions, which are typical of models encountered in game theory, multicriteria models of optimal planning and operations research. Such problems arise also in the solution of systems of equations and inequalities and in nonlinear regression problems, if one uses Chebyshev's cost criterion given by the maximum of the residuals.

Thirdly, nonlinear programming problems may be solved by the method of nonsmooth penalty functions. Certain types of nonsmooth penalty functions are superior to the commonly used smooth penalty functions, because usually they do not require penalty coefficients to tend to infinity, yielding exact solutions for sufficiently large, but finite, values of these coefficients [22, 9].

Fourthly, one may mention optimal control problems with continuous or discrete time. The application of the maximum principle or the discrete maximum principle leads in many cases to problems of minimizing functions with discontinuous gradients. Optimal control problems may be also considered as special nonlinear programming problems which can be solved by decomposition methods or the method of nonsmooth penalty functions, and this again yields nonsmooth optimization problems.

Fifthly, integer and mixed integer programming problems may be indicated. Many such problems can be rather successfully solved by means of the branch and bound method with the bounds resulting from the solution of dual problems. The dual problem usually consists in minimizing, subject to simple constraints, a convex piecewise linear function with an enormous number of pieces, i.e. the dual is a nonsmooth optimization problem.

Finally, functions with discontinuous gradients may directly enter models of optimal planning, design or operations research as a result of piecewise smooth approximation of real technological and economic characteristics of the modelled processes.

It is also worth mentioning that in practice no sharp distinction exists between nonsmooth and smooth functions. From the point of view of applied mathematics and computational practice a function with a rapidly changing gradient is similar in its properties to a nonsmooth function. Therefore one may expect that the computational methods designed for nonsmooth problems will also efficiently minimize ill-behaved smooth functions (e.g. gully-shaped functions).

Consequently, the area of applications of methods for nondifferentiable optimization is vast. Elaboration of efficient algorithms for nonsmooth optimization is a key to the numerical solution of many mathematical programming problems, especially those of large dimension.

Computational methods of nonsmooth optimization were developed in several directions: (a) research aimed at solving particular types of minimization problems with nonsmooth functions that have a special, explicitly defined structure; (b) research on the elaboration of algorithms for solving rather general classes of problems, which do not assume in advance the knowledge of the specific structure of the minimized function, but require only the evaluation of the function and its gradients (or their analogues in the nondifferentiable case) at any given point.

To the first group there belong numerous works on algorithms for minimizing max functions [18]. The existing algorithms for solving minimax problems are essentially close to the methods of feasible directions, being monotone in objective value and requiring at each iteration the solution of auxiliary linear or quadratic programming problems.

A number of works are devoted to the minimization of piecewise linear, convex functions [41]. In this case the techniques of linear programming are usually applied.

For solving several specific nonsmooth optimization problems in which the surfaces of discontinuity of the gradients are given explicitly (e.g. functions involving absolute values), special smoothing techniques have been developed, which enable one to solve such problems approximately by methods for minimizing smooth functions [10].

For solving systems of linear or convex nonlinear inequalities, methods of Fejer-type approximations are used [21, 23], which extend the idea of relaxation methods of Agmon and Motzkin [2].

As far as general methods of nonsmooth optimization are concerned, leaving out such generally applicable methods as the algorithms of random or systematic search, one may distinguish two basic classes, both of which require the computation of generalized gradients.

1. The cutting plane method for solving convex programming problems. The two best known algorithms are the following:
(a) Kelley's cutting plane method [48], based on piecewise linear approximations of the graph of a convex function by supporting hyperplanes; at each iteration of the method one solves a linear programming problem with an increasing number of constraints.
(b) The method of centered sections of A. Yu. Levin, using supporting hyperplanes to the level sets of a convex function at the center of gravity of the area locating the minimum [53]. Such successive cut-offs prove to be rather efficient, but the complexity of the computation of the center of gravity in the multidimensional case impairs the practical value of the method (cf. also [103, 98]). Similar procedures, in which instead of the center of gravity other characteristic points of convex bodies are considered, e.g. the Chebyshev center, were studied by M. E. Primak [67].

2. The generalized gradient methods, which are the major topic of this monograph.

For minimizing smooth functions manifold versions of gradient descent are frequently used, which is natural, since the direction negative to that of the gradient at a given point is a local direction of steepest descent. (Modern gradient methods for minimizing smooth functions are discussed in [54, 69].) Stepsize selection in most of these methods aims at achieving a monotone and significant decrease in the objective function value at each iteration.

An attempt to extend these methods to functions with discontinuous gradients encounters at least two difficulties. First, one has to define an analogue of the gradient at the points at which it does not exist. Secondly, one has to develop new ways of choosing search directions and steplengths, since for nonsmooth functions a straightforward application of the techniques developed for smooth functions can hardly be successful.

The first difficulty can be overcome rather easily. Indeed, in most applications we may confine ourselves to the class of almost everywhere differentiable functions, for which an extension of the gradient at points of nondifferentiability can be defined by taking limits. In fact, the limit can be non-unique, and then instead of a generalized gradient we obtain a gradient set. For a convenient application of the techniques of convex analysis, these sets are enlarged by convexification. In this way generalized gradient sets are obtained, which in the special case of convex functions are called subdifferentials. Minimization algorithms, however, use only arbitrary elements of generalized gradient sets. As will be shown below, in most applications the calculation of generalized gradients is in no way more complex than the computation of gradients; moreover, for almost everywhere differentiable functions the set of points of nondifferentiability is of measure zero.

It is much more difficult, in the case of nonsmooth optimization problems, to find reasonable search directions and steplengths. Even for a convex function the direction negative to that of an arbitrary subgradient is not always a direction of descent. As a matter of fact, with a sufficiently small stepsize the distance to a minimum point decreases, but this distance cannot be calculated in the course of computation, and hence it cannot be used for stepsize selection. Another stepsize rule is needed. The simplest generalized method consists in taking at each iteration a step in the direction negative to that of a generalized gradient. This method, under the name of the "generalized gradient descent", was proposed by the author in 1961, in connection with the need for an efficient algorithm for solving large scale transportation problems, that arose in the problems of operative planning analyzed by the Institute of Cybernetics of the Ukrainian Academy of Sciences in cooperation with the National Planning Council of the Ukrainian Soviet Socialist Republic. Initially the generalized gradient method for minimizing piecewise linear convex functions was applied to the transportation and production problems [75]. Next, this method (known in the West as the subgradient algorithm) was extended to the class of arbitrary convex functions [76] and convex programming problems in Hilbert spaces [64]. Stochastic analogues of the subgradient method have also been extensively developed [25]. Algorithms for solving mathematical programming problems which are based on the idea of the subgradient method, are distinguished by their simplicity and efficient use of the memory of a computer, which is especially important for large scale problems.

The drawbacks of the subgradient method are its rather slow convergence and the complexity of assessing the accuracy of the approximate solution. Moreover, the method is limited to the class of convex functions. All the same, until 1970 considerable experience was gained in the successful solution of many important, practical large scale problems by the subgradient method.

In 1969−1970 the author proposed accelerated versions of the subgradient method, which are based on an application of the operation of space dilation in the direction of the gradient or of the difference of two successive gradients [82]. The concept of these methods differs fundamentally from the one used for accelerating convergence in the case of smooth functions, i.e. from the idea of a quadratic approximation of the function in a neighborhood of a minimum, which yields, in a sense, the formalism of the conjugate gradient method and quasi-Newton methods [45]. At the same time some extreme versions of the methods with space dilation have a quadratic rate of convergence, under certain conditions of smoothness and regularity. Therefore, the algorithms proposed are also highly efficient for smooth minimization problems. The algorithms with space dilation were

further extended to problems of finding local minima of nonconvex an
nonsmooth functions [84].

In the West serious work on gradient methods for minimizing noi
smooth functions started about the year 1973, initially in connection wit
the applications in the area of integer programming [43], and next f(
solving large scale problems [44]. The results of the research conducted i
the West in this direction are presented in the study [3]. A particularly intensiv
development took place in the so-called $\varepsilon$-subgradient optimization, whic
is, on the one hand, close in spirit to the algorithms of V. F. Demyanov f(
solving minimax problems, and on the other hand, formally analogous i
the methods of conjugate gradients (or Davidon's variable metric methods)

Finally, interesting relations have been recently discovered between th
algorithms of successive cut-offs and the algorithms with space dilatic
[103, 87].

To sum up, the research in the area of generalized gradient methods f(
nonsmooth optimization is not yet completed; on the contrary, it is deve
oping intensively.

# 1. Special Classes of Nondifferentiable Functions and Generalizations of the Concept of the Gradient

## 1.1 The Need to Introduce Special Classes of Nondifferentiable Functions

When solving many mathematical programming problems one has to consider a broader class of functions than that of continuously differentiable functions. In such cases we, as a rule, deal with continuous and almost everywhere differentiable functions, whose gradient is discontinuous on a set of null measure. Although in principle functions may arise with even greater pathology (for example, functions that are nowhere differentiable on a set of positive measure), such functions are not encountered in practice; moreover, construction of any nontrivial algorithm for minimizing such functions does not appear possible.

We shall consider such special classes of nondifferentiable functions which, on the one hand, do not generalize too broadly the class of continuously differentiable functions in the sense that for functions from these classes one may reasonably define the generalized notion of the gradient and then use it for constructing gradient-type processes for finding extrema. On the other hand, the special classes of functions to be introduced should contain functions encountered in most applications where one deals with continuous functions defined on $n$-dimensional Euclidean space $E_n$.

The most extensively studied classes are those of convex and concave functions, which are of particular interest from both the point of view of the theory and that of applications of mathematical programming. In recent years various extensions of the class of convex functions have been proposed [61, 68, 56]. In mathematical analysis various classes of piecewise smooth functions have been studied in detail. The class of almost differentiable functions, introduced by the author, may be considered as an extension of the class of continuous and piecewise continuously differentiable functions which satisfy a local Lipschitz property [84]. The class of locally Lipschitzian functions, studied by F. H. Clarke [14], is close to the class of almost differentiable functions. Several interesting classes of nonsmooth functions have been introduced by R. Mifflin [56].

For the classes of functions mentioned above, generalized concepts of gradients (gradient sets) have been introduced either by making use of separability theorems (in this manner subgradients are defined in the convex case), or via the process of taking limits.

## 1.2 Convex Functions. The Concept of Subgradient

We shall review a number of fundamental properties of convex sets and functions which will be useful in what follows.

A subset $W$ of a linear space $L$ is *convex* if $x \in W$ and $y \in W$ implies that $(1 - \alpha) x + \alpha y \in W$ for any $\alpha \in [0, 1]$.

A hyperplane $P$ passing through a boundary point $x$ of a convex set $W \subseteq E_n$ is called a *supporting hyperplane* of $W$ at $x$, if $W$ lies in one of the two closed halfspaces which have $P$ as their common boundary.

**Theorem 1.1.** *At any boundary point of a convex set $W \subseteq E_n$ there exists at least one supporting hyperplane.*

Convex sets possess so-called 'separability properties. Some of them are formulated below.

**Theorem 1.2.** *Let $W_1$ and $W_2$ be closed, disjoint and convex sets in $E_n$, at least one of them bounded. Then there exists a hyperplane*

$$P = \{x: (a, x) + b = 0\}$$

*such that*

$$(a, x) + b < 0 \quad \text{for all } x \in W_1,$$
$$(a, x) + b > 0 \quad \text{for all } x \in W_2.$$

**Theorem 1.3.** *Let $W_1, W_2 \subseteq E_n$ be convex and closed sets with disjoint interiors and assume that the interior of $W_1$ is nonempty. Then for any $\bar{x} \in W_1 \cap W_2$ there exists a hyperplane $(a, x - \bar{x}) = 0$, $a \neq 0$, such that*

$$(a, x - \bar{x}) \leq 0 \quad \text{for all } x \in W_1,$$
$$(a, x - \bar{x}) \geq 0 \quad \text{for all } x \in W_2.$$

The smallest affine set containing a convex set $W$ is called the *affine span* of $W$ and denoted by $N(W)$. The *dimension* $\varrho(W)$ of a set $W$ is defined as the dimension of the manifold $N(W)$. A point $x \in W$ is called *$N$-interior* if it has a neighborhood $S(x)$ such that $S(x) \cap N(W) \subseteq W$. Otherwise we say that $x \in W$ is *$N$-boundary*. A point $x \in W$ is called *extremal* if it cannot be expressed as:

$$x = \alpha x_1 + (1 - \alpha) x_2; \quad 0 < \alpha < 1; \quad x_1, x_2 \in W; \quad x_1, x_2 \neq x.$$

Every extremal point is *$N$-boundary*.

**Theorem 1.4.** *Any point $x$ of a convex, closed and bounded set $W$ can be represented as a convex combination of at most $\varrho(W) + 1$ extremal points of $W$, where $\varrho(W)$ is the dimension of $W$, i.e.*

$$x = \sum_{i=1}^{\varrho(W)+1} \alpha_i x_i, \quad \sum_{i=1}^{\varrho(W)+1} \alpha_i = 1, \quad \alpha_i \geqq 0, \ x_i - \text{extremal points of } W,$$
$$i = 1, \ldots, \varrho(W) + 1.$$

Let $f(x) = f(t_1, \ldots, t_n)$ be a real-valued function defined on a set $M \subseteq E_n$. Consider the $(n+1)$-dimensional space of points of the form $z = \{u, t_1, \ldots, t_n\} = \{u, x\}$. The set $M_f = \{z : u \geqq f(x), x \in M\}$ is called the *epigraph* of the function $f$.

A function $f$ defined on a set $M$ is *convex* if its epigraph is a convex set. A function $\varphi$ is said to be *concave* if the function $-\varphi$ is convex.

**Theorem 1.5.** *A function $f$ is convex if and only if its domain $M$ is convex and for all $x_1, x_2 \in M$ and $\alpha \in [0, 1]$ the following inequality holds:*

$$(1 - \alpha) f(x_1) + \alpha f(x_2) \geqq f((1 - \alpha) x_1 + \alpha x_2).$$

**Theorem 1.6.** *A convex function is continuous at all interior points of its domain.*

Let us now pass on to the definition of the notion of subgradient (generalized gradient). As follows from Theorem 1.1, there exists at any boundary point of a convex set at least one supporting hyperplane. Let us apply this property to the epigraph $M_f$ of the convex function $f$. Any point $z$ of the graph of this function (i.e. $z = \{u, x\}$, $u = f(x)$), is a boundary point of the epigraph.

Let $x_0$ be an interior point of $M$ and let $z_0 = \{f(x_0), x_0\}$. Let us write the equation of the supporting hyperplane of $M$ at $z_0$:

$$(a, z - z_0) = 0, \quad a \neq 0, \tag{1.1}$$

where $a$ is an $(n+1)$-vector. For ease of subsequent arguments, we denote by $\lambda$ the first component of $a$: $a = \{\lambda, b\}$, where $b$ is an $n$-vector. Then (1.1) may be written as

$$\lambda(u - f(x_0)) + (b, x - x_0) = 0.$$

Let the normal vector $a$ of the supporting hyperplane be outer with respect to the set $M_f$. Then for any $z = \{u, x\} \in M_f$ we have

$$\lambda(u - f(x_0)) + (b, x - x_0) \leqq 0. \tag{1.2}$$

By the definition of $M_f$, if $\{\bar{u}, x\} \in M_f$ then also $\{u, x\} \in M_f$ for any $u \geqq \bar{u}$. Therefore, inequality (1.2) should hold for arbitrarily large $u$. Hence, $\lambda \leqq 0$.

If we suppose that $\lambda = 0$ then, taking into account that $x_0$ is an interior point of $M$ and that $(b, x - x_0) \leqq 0$ for all $x \in S(x_0)$, where $S(x_0)$ is a sufficiently small neighbourhood of $x_0$, we obtain $b = 0$. But then $a = \{\lambda, b\} = 0$, which contradicts (1.1). Therefore, $\lambda < 0$.

Setting $g(x_0) = -b/\lambda$ and $u = f(x)$, we obtain

$$f(x) - f(x_0) \geqq (g(x_0), x - x_0) \quad \text{for all } x \in M. \tag{1.3}$$

Hence at any interior point $x_0$ of the domain of a convex function there exists a vector $a(x_0)$ satisfying inequality (1.3).

**Definition.** *Let $f$ be a convex function with domain $M$ and let $x_0$ be an interior point of $M$. A vector $g(x_0)$ satisfying (1.3) for any $x \in M$ is called a subgradient or a generalized gradient of $f$ at $x_0$.*

**Theorem 1.7.** *The set of generalized gradients (the subdifferential) $G(x_0)$ of a convex function $f$ at any interior point $x_0$ of the domain $M$ is nonempty, bounded, convex and closed.*

*Proof.* The existence of at least one subgradient follows from the preceding argument. We shall demonstrate the boundedness of $G(x_0)$. Let $G(x_0)$ contain nonzero points. Consider

$$x(\varepsilon, g) = x_0 + \frac{\varepsilon g}{\|g\|}. \tag{1.4}$$

where

$$\varepsilon > 0, \quad g \in G(x_0), \quad g \neq 0.$$

For any sufficiently small $\varepsilon > 0$ we have $x(\varepsilon, g) \in M$. It follows from the continuity of $f$ at $x_0$ that for any $\delta > 0$ there exists $\varepsilon > 0$ such that

$$|f(x) - f(x_0)| < \varepsilon \quad \text{if} \quad \|x - x_0\| \leqq \delta.$$

On the other hand, by (1.3) and (1.4), $f(x) - f(x_0) \geqq \varepsilon \|g\|$, hence $\|g\| < \delta/\varepsilon$, i.e. $G(x_0)$ is bounded.

The closedness of the set $G(x_0)$ follows from the fact that inequality in (1.3) is preserved when one passes to the limit. It is also easy to verify the convexity. Indeed, let $g_1$ and $g_2$ be subgradients at $x_0$. Then for any $x \in M$

$$f(x) - f(x_0) \geqq (g_1, x - x_0),$$
$$f(x) - f(x_0) \geqq (g_2, x - x_0).$$

Let $\alpha_1, \alpha_2 \geqq 0$, $\alpha_1 + \alpha_2 = 1$. Multiply the first inequality by $\alpha_1$ and the second one by $\alpha_2$ and then add the results:

$$(\alpha_1 + \alpha_2)[f(x) - f(x_0)] = f(x) - f(x_0) \geqq (\alpha_1 g_1 + \alpha_2 g_2, x - x_0),$$
$$\text{for all } x \in M.$$

This yields that $g_3 = \alpha_1 g_1 + \alpha_2 g_2$ is a subgradient, i.e. a convex combination of subgradients is a subgradient. Therefore, $G(x_0)$ is convex. $\square$

**Theorem 1.8.** *A convex function $f$ has a (one-sided) directional derivative in any direction $\eta$ at any interior point $x_0$ of its domain $M$. This derivative satisfies*

$$f'_\eta(x_0) = \max_{g \in G(x_0)} (g, \eta). \tag{1.5}$$

*Proof.* Consider the function

$$\varphi(t) = \frac{f(x_0 + t\eta) - f(x_0)}{t}.$$

Let $\varepsilon > 0$ be sufficiently small so that $x_0 \pm \varepsilon\eta \in M$. Then the function $\varphi$ is continuous on the interval $0 < t < \varepsilon$. We shall prove that it is nondecreasing. Let $0 < t_1 < t_2 < \varepsilon$. We obtain

$$\varphi(t_2) - \varphi(t_1) = \frac{1}{t_1 t_2} [t_1 f(x_0 + t_2\eta) - t_2 f(x_0 + t_1\eta) + (t_2 - t_1)f(x_0)]$$

$$= \frac{1}{t_1} \left\{ \left[ \frac{t_1}{t_2} f(x_0 + t_2\eta) + \left(1 - \frac{t_1}{t_2}\right) f(x_0) \right] - f\left[ \frac{t_1}{t_2}(x_0 + t_2\eta) + \left(1 - \frac{t_1}{t_2}\right) x_0 \right] \right\}.$$

It follows from Theorem 1.5 that $\varphi(t_2) - \varphi(t_1) \geq 0$. We shall prove that $\varphi$ is bounded from below on the interval $(0, \varepsilon)$. To this end consider $\varphi(-\varepsilon/2)$. It is easy to observe that $\varphi(-\varepsilon/2)$ is a lower bound for $\varphi(t)$ for $0 < t < \varepsilon$. Therefore the limit $\lim_{t \to 0+} \varphi(t) = f'_\eta(x_0)$ exists, and $f'_\eta(x_0) \leq \varphi(t)$ for $t \geq 0$.

We shall now prove formula (1.5). For any $g \in G(x_0)$ and $t > 0$ the following inequality holds:

$$\varphi(t) = \frac{f(x_0 + t\eta) - f(x_0)}{t} \geq \frac{(g, t\eta)}{t} = (g, \eta).$$

Thus

$$f'_\eta(x_0) \geq \max_{g \in G(x_0)} (g, \eta). \tag{1.6}$$

Let us consider, in $(n+1)$-dimensional space, the epigraph of the function $f$

$$M_f = \{(u, x) : u \geq f(x)\},$$

and the ray

$$L_\eta = \{(u, x) : x = x_0 + t\eta, \, u = f(x_0) + f'_\eta(x_0)\, t, \, t \geq 0\}.$$

Since $f'_\eta(x_0) \leq \varphi(t)$ for all $t \geq 0$, this ray does not pass through interior points of the epigraph. This, together with Theorem 1.3, implies that there exists a hyperplane supporting the epigraph at the point $\{f(x_0), x_0\}$ and defined by a subgradient $\bar{g}(x_0)$ satisfying:

$$f(x) - f(x_0) \geq (\bar{g}(x_0), x - x_0) \geq t f'_\eta(x_0) \quad \text{for} \quad x = x_0 + t\eta.$$

Thus $(\bar{g}(x_0), \eta) \geq f'_\eta(x_0)$. Combining this inequality with (1.6) we obtain $f'_\eta(x_0) = \max_{g \in G(x_0)} (g, \eta)$, as required. $\quad\square$

**Theorem 1.9.** *A function $f$ is convex on $E_n$ if and only if its directional derivative exists at all points and $f'_\eta(x+t\eta)$ is a nondecreasing function of $t$ for any $x$ and $\eta$.*

*Proof.* The necessity follows directly from Theorem 1.8. Let $t_1 < t_2$. Then

$$f'_\eta(x+t_1\eta) = \max_{g \in G(x+t_1\eta)} (g,\eta) = (g_1^*,\eta), \quad g_1^* \in G(x+t_1\eta);$$

$$f'_\eta(x+t_2\eta) = \max_{g \in G(x+t_2\eta)} (g,\eta) = (g_2^*,\eta), \quad g_2^* \in G(x+t_2\eta);$$

$$(t_2-t_1)f'_\eta(x+t_2\eta) = (t_2-t_1)(g_2^*,\eta) \geqq f(x+t_2\eta) - f(x+t_1\eta)$$
$$\geqq (g_1^*,(t_2-t_1)\eta) = (t_2-t_1)(g_1^*,\eta)$$
$$= (t_2-t_1)f'_\eta(x+t_1\eta).$$

Hence $f'_\eta(x+t_2\eta) \geqq f'_\eta(x+t_1\eta)$.

To demonstrate the sufficiency let us observe that for $t_2 > t_1$, since $f'_\eta(x+t\eta)$ does not decrease in $t$, we have

$$f(x+t_2\eta) - f\left(x + \frac{t_1+t_2}{2}\eta\right) = \int_{(t_1+t_2)/2}^{t_2} f'_\eta(x+t\eta)\,dt$$

$$\geqq \int_{t_1}^{(t_1+t_2)/2} f'(x+t\eta)\,dt = f\left(x + \frac{t_1+t_2}{2}\eta\right) - f(x+t_1\eta).$$

Thus

$$\tfrac{1}{2}[f(x+t_1\eta) + f(x+t_2\eta)] \geqq f\left(x + \frac{t_1+t_2}{2}\eta\right),$$

i.e. $f$ is convex. This proves the result. $\square$

**Theorem 1.10.** *If a function $f$ is twice continuously differentiable, then it is convex if and only if its Hessian $H(x)$ is a positive semidefinite matrix at any point $x$.*

*Proof.* Let us use Taylor's expansion

$$f(x) = f(x_0) + (g(x_0), x-x_0) + \tfrac{1}{2}(H(x_0)(x-x_0), x-x_0) + o(\|x-x_0\|^2)$$

to compute the directional derivatives

$$f'_\eta(x_0+t\eta) = (g(x_0),\eta) + t(H(x_0)\eta,\eta) + o(t),$$
$$f''_\eta(x_0) = (H(x_0)\eta,\eta).$$

The Hessian is positive semidefinite if and only if the second directional derivative is nonnegative for all $x$ and $\eta$, which in turn is equivalent to $f'_\eta(x_0+t\eta)$ being a nondecreasing function of $t$. Applying Theorem 1.9 we obtain the desired result. $\square$

A direction $\eta$, $\eta \neq 0$, will be called a *direction of steepest descent* of a convex function $f$ at an interior point $x_0$ of the domain $M$, if $\min\limits_{\|\xi\|=1} f'_\xi(x_0) = \dfrac{1}{\|\eta\|} f'_\eta(x_0)$.

**Theorem 1.11.** *Suppose $0 \notin G(x_0)$ and let $\eta$ be the element of $G(x_0)$ that is nearest to the origin. Then $\eta$ is a direction of steepest descent at $x_0$.*

*Proof.* Consider the ball $S = \{x : \|x\| \leq \|\eta\|\}$, where $\eta$ is the common boundary point of $S$ and $-G(x_0)$. By virtue of the separation theorem 1.3, the point $\eta$ is contained in a separating hyperplane which supports both $S$ and $-G(x_0)$. Since $S$ is a ball, the only hyperplane supporting $S$ at $\eta$ is given by $(\eta, x - \eta) = 0$, and so $(\eta, x - \eta) \leq 0$ for all $x \in S$ and $(\eta, x - \eta) \geq 0$ for all $x \in -G(x_0)$. Thus $(\eta, x) \geq \|\eta\|^2$ for all $x \in -G(x_0)$ and

$$f'_\eta(x_0) = \max_{g \in G(x_0)} (\eta, g) = - \min_{g \in -G(x^0)} (\eta, g) = -(\eta, \eta).$$

Let $\|\xi\| = 1$. Then

$$f'_\xi(x_0) = \max_{g \in G(x_0)} (g, \xi) = - \min_{g \in -G(x_0)} (g, \xi)$$

$$\geq - \min_{g \in -G(x^0)} \|g\| = - \|\eta\| = \frac{f'_\eta(x_0)}{\|\eta\|}.$$

Consequently $\min\limits_{\|\xi\|=1} f'_\xi(x_0) = \dfrac{1}{\|\eta\|} f'_\eta(x_0)$, i.e. $\eta$ is a direction of steepest descent, which was what we set out to prove. $\square$

If the function is continuously differentiable then $-\eta$ is equal to the gradient.

**Corollary.** *For an interior point $x_0$ of the domain $M$ of $f$ to be a point of the minimum, it is necessary and sufficient that $0 \in G(x_0)$.*

Indeed, if $0 \in G(x_0)$ then for any $x \in M$ one has $f(x) - f(x_0) \geq 0$ and $x_0$ is a minimum point of $f$. If $0 \notin G(x_0)$, then the direction of steepest descent exists, and along it the directional derivative is negative, i.e. $f$ does not attain is minimum at $x_0$.

If the domain $M \subseteq E_n$ has dimension smaller than $n$, then the $N$-interior points play the role of interior points. By analogy with the subdifferential, one can introduce the notion of *N-subdifferential* (*relative subdifferential*) $NG(x_0)$ as the set of vectors $g$ that satisfy the condition $f(x) - f(x_0) \geq (g, x - x_0)$ for all $x \in M$ and belong to the subspace obtained by the parallel shift of the affine span $N(M)$. The relative subdifferentials $NG(x_0)$ possess the same properties as the ordinary subdifferentials.

## 1.3 Some Methods for Computing Subgradients

We shall consider basic operations with respect to which the class of convex functions is closed: 1) the operation of forming linear combinations of functions with nonnegative coefficients; 2) the operation of taking the pointwise maximum over a finite family of functions. In applications, complicated convex functions are usually obtained from simpler ones (e.g. linear or quadratic) by the superposition of the two indicated operations; the operation of taking the maximum is the main factor in the origination of nonsmooth convex functions from smooth ones.

We shall formulate theorems describing the subdifferentials of functions which are linear combinations of convex functions with nonnegative coefficients, and the subdifferentials of functions which are pointwise maxima (max functions).

**Theorem 1.12.** *Let* $f(x) = \sum_{i=1}^{k} a_i f_i(x)$, *where* $a_i \geqq 0$ *and the functions* $f_i$ *are convex,* $i = 1, \ldots, k$. *Then* $f$ *is convex and its subdifferential at any point* $x_0$ *consists of all the vectors of the form*

$$g(x_0) = \sum_{i=1}^{k} a_i g_i(x_0), \quad g_i(x_0) \in G_{f_i}(x). \tag{1.7}$$

*Proof.* Let $x_1, x_2 \in E_n$, $\alpha_1, \alpha_2 \geqq 0$, $\alpha_1 + \alpha_2 = 1$. Then

$$
\begin{aligned}
f(\alpha_1 x_1 + \alpha_2 x_2) &= \sum_{i=1}^{k} a_i f_i(\alpha_1 x_1 + \alpha_2 x_2) \\
&\leqq \sum_{i=1}^{k} a_i [\alpha_1 f_i(x_1) + \alpha_2 f_i(x_2)] \\
&= \alpha_1 \sum_{i=1}^{k} a_i f_i(x_1) + \alpha_2 \sum_{i=1}^{k} a_i f_i(x_2) \\
&= \alpha_1 f(x_1) + \alpha_2 f(x_2).
\end{aligned}
$$

Hence $f$ is convex. It follows readily from the definition of the directional derivative that

$$f'_\eta(x) = \sum_{i=1}^{k} a_i f'_{i\,\eta}(x). \tag{1.8}$$

Consider any vector $g(x_0)$ of the form (1.7). We obtain

$$
\begin{aligned}
f(x) - f(x_0) &= \sum_{i=1}^{k} a_i [f_i(x) - f_i(x_0)] \geqq \sum_{i=1}^{k} a_i (g_i(x_0), x - x_0) \\
&= \left( \sum_{i=1}^{k} a_i g_i(x_0), x - x_0 \right).
\end{aligned}
$$

Therefore $g(x_0)$ is a subgradient of $f$ at $x_0$. Conversely, let $\hat{g}$ be an extremal point of the set $G_f(x_0)$ and let $\eta$ be the outer normal vector of the hyperplane supporting $G_f(x_0)$ at $\hat{g}$ and having $\hat{g}$ as the only common point with $G_f(x_0)$. Then $f'_\eta(x_0) = \max\limits_{g \in G_f(x_0)} (g, \eta)$ and the maximum is attained only at $g$. But it follows from (1.8) that $f'_\eta(x_0) = (\bar{g}, \eta)$ for $\bar{g} = \sum_{i=1}^k a_i \bar{g}_i(x_0)$, where $\bar{g}_i(x_0)$ is the vector at which $\max\limits_{g \in G_{f_i}(x_0)} (g, \eta)$ is attained. Therefore the subset of extremal points of $G_f(x_0)$ contains only vectors of the form $\sum_{i=1}^k a_i \bar{g}_i(x_0)$, where $\bar{g}_i(x_0) \in G_{f_i}(x_0)$; consequently, any vector from $G_f(x_0)$, being a convex combination of extremal points, is of the same form. This completes the proof. $\square$

**Theorem 1.13.** *Let $f_i$ be convex functions, $i = 1, 2, \ldots, m$. Then the function $\varphi(x) = \max\limits_{i \in \overline{1,m}} f_i(x)$ is convex and the subdifferential of $\varphi$ at a point $x_0$ contains the subgradients $g_{f_i}(x_0)$, $i \in I(x_0)$, where $I(x_0) = \{i : \varphi(x_0) = f_i(x_0)\}$.*

*Proof.* Let $\alpha_1, \alpha_2 \geq 0$, $\alpha_1 + \alpha_2 = 1$, $x_1, x_2 \in E_n$. Then

$$\varphi(\alpha_1 x_1 + \alpha_2 x_2) = f_{i*}(\alpha_1 x_1 + \alpha_2 x_2) \leq \alpha_1 f_{i*}(x_1) + \alpha_2 f_{i*}(x_2)$$
$$\leq \alpha_1 \varphi(x_1) + \alpha_2 \varphi(x_2); \quad i^* \in I(\alpha_1 x_1 + \alpha_2 x_2).$$

This proves the convexity of $\varphi$.

Next, if $i \in I(x_0)$ then

$$(g_{f_i}(x_0), x - x_0) \leq f_i(x) - f_i(x_0) \leq \varphi(x) - \varphi(x_0).$$

Therefore $g_{f_i}(x_0)$ is a subgradient of $\varphi$ at $x_0$, as required. $\square$

Summing up, in many cases important in practice the computation of a subgradient does not essentially differ in complexity from the computation of the gradient. It is also worth noting that a convex function is differentiable almost everywhere (this means that almost everywhere, i.e. except for a set of null measure, the subgradient is identical with the gradient). Moreover, the gradient of a convex function is continuous on the set of points at which it exists [12].

The monotonicity of the directional derivative of a convex function may be exploited in the construction of an algorithm for calculating the directional derivative $f'_\eta(x_0)$ with any required accuracy, provided $f$ is differentiable at $x_0$ and can be exactly evaluated at each point. Indeed, for any $h > 0$ we have

$$\frac{1}{h}[f(x + h\eta) - f(x)] \geq f'_\eta(x) \geq \frac{1}{h}[f(x) - f(x - h\eta)],$$

which yields both upper and lower bounds for the derivative; it is easy to choose $h$ so that the error is within the prespecified limits. Since the calculation of the gradient consists in the evaluation of partial derivatives, we thus obtain an algorithm for computing the gradient with a given accuracy at any point at which the gradient exists.

On the other hand, one can easily construct examples showing that if the subdifferential at a given point is not a singleton, then no algorithm can guarantee for each $\varepsilon > 0$ the computation of a vector $g_\varepsilon$ satisfying

$$\min_{g \in G_f(x)} \| g_\varepsilon - g \| < \varepsilon.$$

However, we can construct an appropriate algorithm for the following relaxed version of the problem of computing the approximation to a subgradient.

Given positive $\varepsilon$ and $\delta$, we shall construct a vector $g^*$ such that in the $\delta$-neighborhood of $x_0$ one can find a point $x$ for which

$$\min_{g \in G_f(x)} \| g^* - g \| < \varepsilon.$$

To motivate the construction we shall use the following fact resulting from the continuity of $f$: if the inequality

$$f(x + h\eta) + f(x - h\eta) - 2f(x) < \bar{\varepsilon} h$$

holds for some $x \in R^n$, $\bar{\varepsilon} > 0$ and $h > 0$, then for any $\bar{\delta} > 0$ ond can find a neighborhood $S_{\bar{\delta}}(x)$ of $x$ such that

$$f(y + h\eta) + f(y - h\eta) - 2f(y) < (\bar{\varepsilon} + \bar{\delta}) h \qquad \text{for all } y \in S_{\bar{\delta}}(x).$$

We shall construct the vectors $g^*$ and $\bar{x}$ in $n$ steps which will generate a sequence of vectors $\bar{x}_k$, $k = 1, 2, \ldots, n$ so that $\bar{x}_n = \bar{x}$.

Let $\{e_1, \ldots, e_n\}$ be an orthonormal basis and let $\bar{x}_k = x_0 + \sum_{i=1}^{k} c_i e_i$. Suppose that $\| \bar{x}_k - x_0 \| \leq \delta k / n$ and that we have $h_k > 0$ such that

$$f(\bar{x}_k + h_k e_i) + f(\bar{x}_k - h_k e_i) - 2f(\bar{x}_k \leq \frac{k \varepsilon h_k}{n \sqrt{n}} \qquad \text{for } 1 \leq i \leq k < n.$$

Consider the points

$$y_r = \bar{x}_k + \frac{\delta r}{pn} e_{k+1}, \qquad -p \leq r \leq p.$$

The sequence $s_r = \dfrac{pn}{\delta} [f(y_{r+1}) - f(y_r)]$, $r = -p, \ldots, p - 1$, is nondecreasing. On the other hand, for any $p$

$$s_{-p} \geq f'_{e_{k+1}}\left(\bar{x}_k - \frac{\delta}{n} e_{k+1}\right); \quad s_{p-1} \leq f'_{e_{k+1}}\left(\bar{x}_k + \frac{\delta}{n} e_{k+1}\right).$$

Combined with the preceding result, this implies that for a sufficiently large $p = \bar{p}$ one can find $j^*$, $-\bar{p} \leq j^* \leq \bar{p} - 1$, such that

$$s_{j^*} - s_{j^*-1} < \frac{(k + 1) \varepsilon}{n \sqrt{n}}$$

and

$$\frac{f(y_{j*} + h_k\, e_i) + f(y_{j*} - h_k\, e_i) - 2f(y_{j*})}{h_k} < \frac{(k+1)\,\varepsilon}{n\,\sqrt{n}}\,.$$

Set $h_{k+1} = \min\left\{h_k,\ \dfrac{\delta}{\bar{p}n}\right\}$ and $\bar{x}_{k+1} = y_{j*}$. Then the following relations hold:

$$f(\bar{x}_{k+1} + h_{k+1}\, e_i) + f(\bar{x}_{k+1} - h_{k+1}\, e_i) - 2f(\bar{x}_{k+1}) < \frac{(k+1)\,\varepsilon\,h_{k+1}}{n\,\sqrt{n}}\,,$$
$$i = 1, \ldots, k+1;$$

$$\|\bar{x}_{k+1} - x_0\| \leq \frac{\delta(k+1)}{n}\,.$$

By induction, after $n$ steps we obtain the vector $\bar{x} = \bar{x}_n$ such that for some $h_n > 0$

$$\frac{1}{h_n}[f(\bar{x} + h_n\, e_i) + f(\bar{x} - h_n\, e_i) - 2f(\bar{x})] \leq \frac{\varepsilon}{\sqrt{n}}\,, \quad i = 1, \ldots, n, \tag{1.9}$$

$$\|\bar{x} - x_0\| < \delta. \tag{1.10}$$

Observe that we have the inequalities:

$$\frac{1}{h_n}[f(\bar{x}) - f(\bar{x} - h_n\, e_i)] \leq f'_{e_i}(\bar{x}) \leq \frac{1}{h_n}[f(\bar{x} + h_n\, e_i) - f(\bar{x})], \tag{1.11}$$

$$\frac{1}{h_n}[f(\bar{x}) - f(\bar{x} - h_n\, e_i)] \leq -f'_{-e_i}(\bar{x}) \leq \frac{1}{h_n}[f(\bar{x} + h_n\, e_i) - f(\bar{x})]. \tag{1.12}$$

Next,

$$f'_{e_i}(\bar{x}) = \max_{g \in G_f(\bar{x})} (g,\, e_i), \tag{1.13}$$

$$-f'_{-e_i}(\bar{x}) = \min_{g \in G_f(\bar{x})} (g,\, e_i). \tag{1.14}$$

Consider the parallelepiped

$$\Pi = \left\{z\colon \frac{f(\bar{x}) - f(\bar{x} - h_n\, e_i)}{h_n} \leq (z,\, e_i) \leq \frac{f(\bar{x} + h_n\, e_i) - f(\bar{x})}{h_n}\right\}.$$

It follows from (1.11) through (1.14) that $G_f(\bar{x}) \subseteq \Pi$. Let us take the center of $\Pi$ as $g*$:

$$g* = \left\{\frac{f(\bar{x} + h_n\, e_i) - f(\bar{x} - h_n\, e_i)}{2h_n}\right\}_{i=1}^{n}.$$

From (1.9) and (1.10) we get

$$\|g* - g\| < \varepsilon/2 \quad \text{for some } g \in G_f(\bar{x}).$$

Thus we have constructed the desired vector $g*$ and the point $\bar{x}$.

The above algorithm can calculate a vector which is arbitrarily close to the subdifferential of $f$ at a point belonging to an arbitrarily small neighborhood of the original point.

## 1.4 Almost Differentiable Functions

Most works devoted to gradient-type methods or generalized gradient methods assume either continuity of the gradient or convexity of the objective function, respectively. But in many practical problems one can hardly postulate smoothness (or convexity) of the objective function. For example, in economic planning some components of the objective function are usually defined as piecewise-smooth and not necessarily convex functions of a parameter which characterizes the productivity (the output capacity) of a certain unit. Nondifferentiability and nonconvexity are typical of a broad class of minimax problems.

Therefore a need arises to consider minimization problems with a class of functions which is broad enough to contain piecewise-smooth functions and functions encountered in minimax problems, and narrow enough to allow a natural generalization of the gradient and an application of gradient-type methods for local minimum seeking.

These requirements are met by the class of almost differentiable functions introduced in [84].

**Definition.** *A function $f$ defined on n-dimensional Euclidean space $E_n$ is called almost differentiable if:*

(a) *in any bounded area it satisfies the Lipschitz condition;*
(b) *it is differentiable almost everywhere;*
(c) *its gradient is continuous on its domain M.*

Note that property (b) is not independent. We have the following theorem [46]:

*Let a real-valued function $f$ defined on an open set $D \subset E_n$ have finite Dini derivatives in all directions, i.e.*

$$\limsup_{t \to 0} \left| \frac{f(x + t v) - f(x)}{t} \right| < + \infty$$

*for any $x \in D$ and any $v \in E_n$. Then $f$ is differentiable almost everyhwere on D.*

It is clear that if a function $f$ satisfies the Lipschitz condition in any open bounded, area $S \subset E_n$, i.e.

$$|f(x) - f(y)| \leqq L \|x - y\|, \quad L > 0; \quad x, y \in S,$$

then

$$\limsup_{t \to 0} \left| \frac{f(x + t v) - f(x)}{t} \right| \leqq L \|v\| < \infty$$

for all $x \in S$. Therefore local Lipschitz continuity implies differentiability almost everywhere.

**Definition.** *An almost-gradient of a function $f$ at a point $x_0$ is an accumulation point of a sequence of gradients $g(x_1), g(x_2), \ldots,$ such that $\{x_k\}_{k=1}^{\infty}$ converges to $x_0$ and $f$ is differentiable at each $x_k$.*

**Theorem 1.14.** *The set $G(x)$ of almost-gradients of an almost differentiable function is at any point $x \in E_n$ nonempty, bounded and closed.*

*Proof.* Property (a) implies the boundedness of the sequence $\{g(x_k)\}$, which in turn implies that $G(x)$ is nonempty and bounded. The closedness follows from the definition of almost-gradients as limit points. Indeed, let $\{\hat{g}^{(k)}(x)\}$ be a sequence of almost-gradients converging to $g^*$. Then for each $k$ one can find $x^{(k)}$ such that $f$ is differentiable at $x^{(k)}$, $\|x - x^{(k)}\| \leq 1/k$ and $\|\hat{g}^{(k)}(x) - g(x^{(k)})\| \leq 1/k$. Then

$$\|g^* - g(x^{(k)})\| \leq \|g^* - \hat{g}^{(k)}(x)\| + \|\hat{g}^{(k)}(x) - g(x^{(k)})\|$$

$$\leq \|g^* - \hat{g}^{(k)}(x)\| + \frac{1}{k}.$$

Since $\lim\limits_{k \to \infty} \|g^* - \hat{g}^{(k)}(x)\| = 0$, it follows that $g^*$ is an almost-gradient of $f$ at $x$. $\quad\square$

**Theorem 1.15.** *A convex function $f$ defined on $E_n$ is almost differentiable and at any point its almost-gradients coincide with subgradients.*

*Proof.* By Rademeister's theorem [12] any convex function defined on $E_n$ is almost everywhere continuously differentiable. It follows from well-known properties of convex functions that directional derivatives of $f$ are uniformly bounded in any bounded area. Therefore $f$ is almost differentiable.

Let $\hat{g}_f(x_0) = \lim\limits_{k \to \infty} g_f(x_k)$ be an almost-gradient, where each $g_f(x_k)$ is the gradient of $f$ at $x_k$ and $\lim\limits_{k \to \infty} x_k = x_0$. It follows from the convexity of $f$ that for all $x \in E_n$

$$f(x) - f(x_k) \geq (g_f(x_k), x - x_k).$$

Letting $k$ approach infinity we obtain

$$f(x) - f(x_0) \geq (\hat{g}_f(x_0), x - x_0),$$

i.e. $\hat{g}_f(x_0)$ is a subgradient of $f$ at $x_0$. This completes the proof. $\quad\square$

**Theorem 1.16.** *If functions $f_1$ and $f_2$ are almost differentiable then also the functions $f_3(x) = f_1(x) \pm f_2(x)$ and $f_4(x) = f_1(x) f_2(x)$ are almost differentiable.*

*Proof.* Define the set $N_3$ as the union of the sets $N_1$ and $N_2$ on which the functions $f_1$ and $f_2$ are nondifferentiable. The measure of $N_3$ equals zero.

Outside $N_3$ both $f_3$ and $f_4$ are differentiable and their gradients are given by classical rules for differentiating the sum and the product of two functions. This proves the theorem.   $\Box$

**Theorem 1.17.** *If $f_1$ and $f_2$ are almost differentiable, then the function $f(x) = \max(f_1(x), f_2(x))$ is almost differentiable.*

*Proof.* Denote

$$\varphi^+(x) = \begin{cases} 0 & \text{if } \varphi(x) \leq 0. \\ \varphi(x) & \text{if } \varphi(x) > 0. \end{cases}$$

Then

$$f(x) = f_1(x) + [f_2(x) - f_1(x)]^+.$$

By Theorem 1.15, $f_2(x) - f_1(x)$ is almost differentiable. Next, by the continuity of $f_1$ and $f_2$, the set $M_{12} = \{x : f_1(x) > f_2(x)\}$ is open. Thus $f$ is differentiable at all those points of $M_{12}$ at which $f_1$ is differentiable. A similar statement holds for the set $M_{21} = \{x : f_1(x) < f_2(x)\}$.

As far as the set $M_0 = \{x : f_1(x) = f_2(x)\}$ is concerned, its boundary is of null measure, and in its interior the differentiability of $f$ results from the differentiability of $f_1$. This completes the proof.   $\Box$

**Definition.** *Any vector in the closure of the convex hull of the set of almost-gradients of $f$ at $x$ is called a generalized almost-gradient of $f$ at $x$.*

It follows from Theorem 1.14 that the set of generalized almost-gradients is convex, bounded and closed.

F. H. Clarke [15] introduced another concept of generalized gradient, which is close to our definition. In his paper the class of locally Lipschitz functions on $E_n$ is considered and the following definition is formulated; the *generalized gradient* of $f$ at $x$, denoted by $\partial f(x)$, is the closure of the convex hull of the set of all limits of sequences of the form $\{g_f(x + h_i)\}_{i=1}^{\infty}$ with $\lim_{i \to \infty} h_i = 0$.

Our definition differs from Clarke's in the additional requirement that the gradient be continuous on its domain.


## 1.5 Semismooth and Semiconvex Functions

Following [56], we shall present basic analytical properties of some important classes of nondifferentiable functions.

In [15] F. H. Clarke considered the class of locally Lipschitz functions and the following analogue of the directional derivative

$$f^0(x, d) = \lim_{\substack{h \to 0 \\ t \to +0}} \sup \frac{1}{t} [f(x + h + t\,d) - f(x + h)].$$

The *generalized gradient set* (called *generalized gradient* by Clarke) of $f$ at $x$ is defined by

$$\partial f(x) = \{g \in E_n : (g, d) \le f^0(x, d) \text{ for all } d \in E_n\}$$

and has the following properties:

(a) $\partial f(x)$ is a nonempty convex compact set;
(b) $f^0(x, d) = \max_{g \in \partial f(x)} (g, d)$;
(c) $f$ is differentiable almost everywhere and $\partial f(x)$ is the closure of the convex hull of all points $g$ of the form $g = \lim_{k \to \infty} g_f(x_k)$, where $\{x_k\}_{k=1}^{\infty} \to x$ and $f$ has the gradient $g_f(x_k)$ at each $x_k$;
(d) if $\{x_k\}$ converges to $x$ and $g_k \in \partial f(x_k)$ for all $k$, then the sequence $\{g_k\}$ is bounded and all its accumulation points belong to $\partial f(x)$, i.e. the point-to-set mapping $x \to \partial f(x)$ is bounded and upper semicontinuous;
(e) an analogue of the mean value theorem holds: for any two points $y$ and $z$ one can find $\lambda \in (0; 1)$ and $g \in \partial f(y + \lambda(z - y))$ such that $f(z) - f(y) = (g, z - y)$.

If the directional derivative

$$f'(x, d) = \lim_{t \to +0} \left[ \frac{1}{t} f(x + t d) - f(x) \right]$$

exists and equals $f^0(x, d)$, then $f$ is called *quasidifferentiable* [68].

R. Mifflin introduced the class of *semismooth functions*.

**Definition.** *A function $f$ defined on $E_n$ is called semismooth at a point $x \in E_n$, if:*

(a) *$f$ is Lipschitz continuous in a neighborhood of $x$;*
(b) *for any $d \in E_n$ and any sequences $\{t_k\}_{k=1}^{\infty} \subset E_1$, $\{\theta_k\}_{k=1}^{\infty} \subset E_n$ and $\{g_k\}_{k=1}^{\infty} \subset E_n$ such that $t_k \to +0$, $\dfrac{\theta_k}{t_k} \to 0$ and $g_k \in \partial f(x + t_k d + \theta_k)$, the sequence $\{g_k, d)\}_{k=1}^{\infty}$ has a unique accumulation point.*

The following theorem holds: if $f$ is semismooth at $x$ then for each $d \in E_n$ the directional derivative $f'(x, d)$ exists and equals $\lim_{k \to \infty} (g_k, d)$, where $\{g_k\}$ is any sequence satisfying the conditions in (b) above.

**Definition.** *A function $f$ is semiconvex at a point $x \in E_n$ if:*

(a) *$f$ is Lipschitz in a neighborhood of $x$;*
(b) *$f$ is quasidifferentiable at $x$;*
(c) *$f'(x, d) \ge 0$ implies $f(x + d) \ge f(x)$.*

A function which is both semiconvex and differentiable is called *pseudoconvex* [70]. We say that a function $f$ is semismooth (quasidifferentiable, semiconvex) on $X \subseteq E_n$ if it is semismooth (quasidifferentiable, semiconvex) at all $x \in X$. For semismooth functions an interesting superposition theorem was proved in [56].

**Theorem 1.18.** *Suppose that functions $f_i$, $i = 1, \ldots, m$, are semismooth on $E_n$, function $\Phi$ is semismooth on $E_m$ and define*

$$y(x) = \{f_1(x), \ldots, f_m(x)\}; \quad F(x) = \Phi(y(x));$$

$$G(x) = \operatorname{conv}\left\{g \in E_n: g = \sum_{i=1}^{m} w_i\, g^{(i)}, (w_1, w_2, \ldots, w_m) \in \partial\Phi(y(x)), \right.$$

$$\left. g^{(i)} \in \partial f_i(x), i = 1, \ldots, m\right\}.$$

*Then the function $F$ is semismooth and $\partial F(x) \subseteq G(x)$.*

Unfortunately, it may happen that $\partial F(x) \neq G(x)$, which complicates the evaluation of generalized gradients in some cases.

For semismooth functions Mifflin obtained interesting generalized gradient-type processes converging to *stationary points,* i.e. points at which the generalized gradient set contains the null vector.

E. A. Nurminski [61] introduced the class of *weakly convex functions.*

**Definition.** *A continuous function $f$ defined on $E_n$ is called weakly convex if for any $x \in E_n$ one can find a nonempty set $M(x)$ of vectors $g$ such that for all $y \in E_n$*

$$f(y) - f(x) \geqq (g, y - x) - r(x, y),$$

*where $r(x, y)/\|x - y\| \to 0$ as $y - x \to 0$, uniformly in all compact subsets of $E_n$.*

Weakly convex functions form a subclass of the family of quasidifferentiable functions. Both classes are closed with respect to the operation of taking the pointwise maximum [61], which motivates their application in the analysis of minimax problems.

# 2. The Subgradient Method

## 2.1 The Problem of Stepsize Selection in the Subgradient Method

Let $f$ be a convex function defined on $E_n$. The *subgradient method* is an algorithm which generates a sequence $\{x_k\}_{k=0}^{\infty}$ according to the formula

$$x_{k+1} = x_k - h_{k+1}(x_k)\, g_f(x_k'),\tag{2.1}$$

where $x_0$ is a given starting point.

At first sight formula (2.1) does not differ from the formulae defining usual gradient methods, especially if we take into account that the subgradient $g_f(x)$ is almost everywhere identical with the gradient, because a convex function is differentiable almost everywhere. But the rules for determining stepsizes $h_{k+1}(x_k)$ must be here entirely different. Indeed, the stepsize rules used for continuously differentiable functions, e.g. $h_{k+1}(x_k) = h$ for some sufficiently small constant $h$, as in the simple gradient method, or the minimization of $f(x_k - h_{k+1}\, g_f(x_k))$ with respect to $h_{k+1}$, as in the method of steepest descent, do not apply to all convex functions. Constant stepsizes are unsuitable because the function may be nondifferentiable at the optimal point and then $\{g_f(x_k)\}_{k=0}^{\infty}$ does not necessarily tend to zero, even if $\{x_k\}_{k=0}^{\infty}$ converges to the optimal point. On the other hand exact directional minimization may cause converge to a nonstationary point, as shown in the example below.

**Example** [101]. Consider the function of two variables

$$f(x_1, x_2) = \begin{cases} 5(9\,x_1^2 + 16\,x_2^2)^{1/2} & \text{for } x_1 > |x_2|, \\ 9\,x_1 + 16\,|x_2| & \text{for } x_1 \leq |x_2|. \end{cases}$$

If $x_1 > 0$ then $f(x_1, x_2) = \max\{5(9\,x_1^2 + 16\,x_2^2)^{1/2}, 9\,x_1 + 16\,|x_2|\}$. Thus $f$ is convex as the pointwise maximum of two convex functions. We shall prove that $f$ is continuously differentiable in this area. The set of possible discontinuities of the gradient satisfies the condition $x_1 = |x_2|$. Let us compute the gradients

$$\partial[5(9\,x_1^2 + 16\,x_2^2)^{1/2}] = \left\{\frac{5\cdot 9\,x_1}{(9\,x_1^2 + 16\,x_2^2)^{1/2}}\,;\ \frac{5\cdot 16\,x_2}{(9\,x_1^2 + 16\,x_2^2)^{1/2}}\right\}\,;$$

$$\partial[5(9\,x_1^2 + 16\,x_2^2)^{1/2}] = \{9;\ 16\ \text{sign}\ x_2\}\quad \text{for}\ x_1 = |x_2|\,;$$

$$\partial[9\,x_1 + 16\,|x_2|]\qquad = \{9;\ 16\ \text{sign}\ x_2\}\,.$$

Thus there is no discontinuity of the gradient on the line $x_1 = |x_2|$, and so $f$ is continuously differentiable on the halfplane $x_1 > 0$. If we choose a starting point in the area $x_1 > |x_2| > (\frac{9}{16})^2\,x_1$ then, as one can easily verify, the method of steepest descent will follow a polygonal path having successive orthogonal segments whose vertices converge to the point $\{0, 0\}$. We know that for continuously differentiable functions the set of accumulation points of a sequence generated by the method of steepest descent consists of stationary points. But in our example the method converges to a nonstationary point $\{0, 0\}$. This results from the discontinuity of the, gradient at that point (the, gradient is discontinuous on the ray $x_2 = 0$, $x_1 \leqq 0$, while $\lim\limits_{x_1 \to -\infty} f(x_1, 0) = -\infty$). The above example shows that the method of steepest descent may be unsuitable for minimizing nondifferentiable convex functions.

Therefore, when the subgradient method is applied to nonsmooth functions, specific difficulties in stepsize selection arise, which makes it necessary to relax the requirement of monotone descent.

To gain an insight into this subject let us consider the subgradient method with constant steplengths, i.e. with

$$h_{k+1}(x_k) = \frac{h}{\|g_f(x_k)\|}$$

where $h > 0$.

**Theorem 2.1.** *Let $f$ be a convex function having a nonempty set of minimum points $M^*$. Suppose that a sequence $\{x_k\}$ is calculated by the subgradient method with stepsizes $h_{k+1}(x_k) = h/\|g_f(x_k)\|$, where $h > 0$. Then for any $\varepsilon > 0$ and any $x^* \in M^*$ one can find $k = k^*$ and $\bar{x}$ such that $f(\bar{x}) = f(x_{k^*})$ and $\|\bar{x} - x^*\| < h(1 + \varepsilon)/2$.*

*Proof.* If $g_f(x_{k^*}) = 0$ for some $k^*$ then $f(x_{k^*}) = f(x^*)$ and one may take $\bar{x} = x^*$. If $g_f(x_k) \neq 0$ for all $k = 0, 1, \ldots$, then

$$x_{k+1} = x_k - \frac{h\,g_f(x_k)}{\|g_f(x_k)\|}\,,\quad k = 0, 1, 2, \ldots .$$

For any $x_k \in E_n$

$$\|x_{k+1} - x^*\|^2 = \|x_k - x^* - h\,g_f(x_k)\|\,g_f(x_k)\|^{-1}\|^2$$

$$= \|x_k - x^*\|^2 + h^2 - 2\,h\left(x_k - x^*,\ \frac{g_f(x_k)}{\|g_f(x_k)\|}\right). \tag{2.2}$$

We shall estimate the term $(x_k - x^*, g_f(x_k)/\|g_f(x_k)\|)$, denoted by $a_k(x^*)$, which is equal to the distance from $x^*$ to the supporting hyperplane $L_k = \{x : (g_f(x_k), x - x_k) = 0\}$. Define $U_k = \{x : f(x) = f(x_k)\}$ and $b_k(x^*) = \varrho(x^*, U_k)$. Since both the set $U_k$ and the point $x^*$ lie on the same side of $L_k$ and any segment joining $x^*$ with a point of $L_k$ passes through $U_k$, we have $a_k(x^*) \geqq b_k(x^*)$.

From (2.2) we obtain

$$\|x_{k+1} - x^*\|^2 \leqq \|x_k - x^*\|^2 + h^2 - 2 h b_k(x^*). \tag{2.3}$$

If $b_k(x^*) \geqq \dfrac{h}{2}(1 + \varepsilon)$ for all $k = 0, 1, 2, \ldots$, then

$$\|x_{k+1} - x^*\|^2 \leqq \|x_k - x^*\|^2 - \varepsilon h^2 \leqq \|x_0 - x^*\|^2 - \varepsilon(k + 1) h^2$$

for all $k$. But $\|x_{k+1} - x^*\|^2 \geqq 0$. Therefore $k^*$ exists such that

$$b_{k^*} = \varrho(x^*, U_{k^*}) < \frac{h}{2}(1 + \varepsilon),$$

as required.   $\square$

From the above theorem we deduce easily the following results.

**Corollary 1.** *Under the assumptions of Theorem 2.1, for any $\delta > 0$ there exists $h_\delta > 0$ such that if the subgradient method is used with stepsizes $h_{k+1}(x_k) = h_\delta/\|g_f(x_k)\|$ then either $x_{k^*} \in M^*$ for some $k^*$ or there is a subsequence $k_1 < k_2 < k_3 < \ldots$ such that*

$$f(x_{k_i}) - f^* < \delta,$$

*where $f^* = \min\limits_{x \in E_n} f(x)$.*

**Corollary 2.** *If the set $M^*$ of minimum points of the function $f$ contains a sphere with a radius $r > h/2 > 0$ and the subgradient method is applied with $h_{k+1}(x_k) = h/\|g_f(x_k)\|$ then there exists $k^*$ such that $x_{k^*} \in M^*$.*


## 2.2  Basic Convergence Results for the Subgradient Method

Let us analyse convergence of several basic versions of the subgradient method. It can be seen from the proof of Theorem 2.1 that at each iteration the reduction in the distance to the set of minima is guaranteed only outside a certain neighborhood of that set, with the size of that neighborhood depending on the value of the steplength $h$. Therefore, to obtain standard convergence theorems it is necessary to require that $h_k$ tends to zero. The reduction of steplengths, however, should not be too rapid. In particular, if the series $\sum_{k=1}^{\infty} h_k$ is convergent then the sequence $\{x_k\}_{k=0}^{\infty}$ has a limit, but this limit may lie outside $M^*$. So we have arrived at the classical conditions: $h_k > 0$, $h_k \to 0$ as $k \to \infty$ and $\sum_{k=1}^{\infty} h_k = +\infty$.

There are many alternative proofs of convergence of the subgradient method [24, 64, 80]. All the proofs are based on the analysis of the behavior of the sequence $\{\varrho_k\}_{k=0}^{\infty}$, where $\varrho_k = \min_{y \in M^*} \| x_k - y \|$. The most general result (for constrained convex problems in Hilbert spaces) was obtained by B. T. Polyak [64]. Yu. M. Ermoliev [24] established convergence in the finite-dimensional case. Below we present our version of the proof of the convergence of the subgradient method.

**Theorem 2.2.** *Let $f$ be a convex function defined on $E_n$ which has a bounded set of minimum points $M^*$ and let a sequence of positive numbers $\{h_k\}$, $k = 1, 2, \ldots$, satisfy the conditions:*

$$\lim_{k \to \infty} h_k = 0, \ \sum_{k=1}^{\infty} h_k = + \infty .$$

*Then for any $x_0 \in E_n$ the sequence $\{x_k\}$, $k = 0, 1, \ldots$, generated according to the formula*

$$x_{k+1} = x_k - h_{k+1} \frac{g_f(x_k)}{\| g_f(x_k) \|} , \tag{2.4}$$

*has the following property: either an index $\bar{k}$ exists such that $x_{\bar{k}} \in M^*$, or $\lim_{k \to \infty} \min_{y \in M^*} \| x_k - y \| = 0$ and $\lim_{k \to \infty} f(x_k) = \min_{x \in E_n} f(x) = f^*$.*

*Proof.* Let $x^* \in M^*$. As in the proof of Theorem 2.1, for $x_k \notin M^*$ we obtain

$$\| x_{k+1} - x^* \|^2 \le \| x_k - x^* \|^2 + h_{k+1}^2 - 2 h_{k+1} b_k (x^*) .$$

For a fixed $a > 0$ consider the set $\{x : f(x) \le f^* + a\}$ and its boundary $U_{f^*+a}$. By assumption, $M^*$ is bounded and closed. Thus $U_{f^*+a}$ is compact. Furthermore, $M^* \cap U_{f^*+a}$ is empty. Hence there exists a number

$$b(a) = \min_{x \in U_{f^*+a}, y \in M^*} \| x - y \| > 0 .$$

Since $\lim_{k \to \infty} h_k = 0$, one find $N_\varepsilon$ such that $h_k < \varepsilon$ for all $k > N_\varepsilon$. Set $\varepsilon = b(a)$. For $k > N_{b(a)}$ we have $h_k < b(a)$. If $f(x_k) \ge f^* + a$ then by (2.3)

$$\| x_k - x^* \|^2 - \| x_{k+1} - x^* \|^2 \ge b(a) h_{k+1} .$$

Since $\sum_k h_k = \infty$, there must exist $N_a^{(1)} > N_{b(a)}$ such that $f(x_{N_a^{(1)}}) \le f^* + a$. Let us define

$$d(a) = \max_{y \in U_{f^*+a}} \min_{x \in M^*} \| y - x \| .$$

If $f(x_k) \le f^* + a$ then $\min_{x \in M^*} \| x_k - x \| \le d(a)$ and $\min_{x \in M^*} \| x_{k+1} - x \| \le d(a) + h_{k+1}$. On the other hand, if $f(x_k) > f^* + a$ and $k > N_{b(a)}$ then

$$\min_{x \in M^*} \| x_{k+1} - x \| \le \min_{x \in M^*} \| x_k - x \| .$$

Hence for all $k > N_a^{(1)}$

$$\min_{x \in M^*} \|x_k - x\| \le d(a) + h(N_a^{(1)}),$$

where $h(N_a^{(1)}) = \max_{k > N_a^{(1)}} h_k$. Since $d(a) \to 0$ as $a \to 0$, for any $\delta > 0$ there exists $a(\delta)$ such that $d(a(\delta)) \le \delta/2$. Next, one can find an index $N_\delta$ such that $f(x_{N_\delta}) \le f^* + a(\delta)$ and $h_k \le \delta/2$ for all $k > N_\delta$. So, for all $k > N_\delta$

$$\min_{y \in M^*} \|x_k - y\| \le \delta.$$

This proves the relation $\lim_{k \to \infty} \min_{y \in M^*} \|x_k - y\| = 0$.

By the continuity of $f$,

$$\lim_{k \to \infty} f(x_k) = f^*,$$

which completes the proof. $\quad \square$

A similar result holds for another stepsize rule which does not use the scaling factors $\|g_f(x_k)\|^{-1}$ for normalizing the subgradients.

**Theorem 2.3.** *Let the assumptions of Theorem 2.2 be satisfied and suppose that a sequence $\{x_k\}_{k=0}^{\infty}$ is generated according to the formula*

$$x_{k+1} = x_k - h_{k+1}\, g_f(x_k), \quad k = 0, 1, \ldots, \tag{2.5}$$

*where $x_0 \in E_n$ is an arbitrary starting point, $h_k > 0$, $h_k \to 0$, and $\sum_{k=1}^{\infty} h_k = \infty$. Then either*

(a) *the sequence $\{g_f(x_k)\}_{k=0}^{\infty}$ is bounded and the algorithm converges in the sense that*

$$\lim_{k \to \infty} \min_{x \in M^*} \|x_k - x\| = 0, \quad \lim_{k \to \infty} f(x_k) = f^*,$$

*or*

(b) *the sequence $\{g_f(x_k)\}_{k=0}^{\infty}$ is unbounded and there is no convergence.*

*Proof.* We shall present a simple example for case (b). Take $f(x) = x^4$, $h_k = 1/k$ and $x_0 = 1$. We obtain $g_f(x_0) = 4$, $x_1 = -3$, $g_f(x_1) = -108$, $x_2 = 51$, etc. It is easy to verify that $\lim_{k \to \infty} |g_f(x_k)| = +\infty$.

Consider case (a). If $g_f(x_{k^*}) = 0$ for some $k^*$ then $x_{k+1} = x_k$ for all $k \ge k^*$ and $x_{k^*} \in M^*$. Suppose that $g_f(x_k) \ne 0$ for all $k$. Define $\bar{h}_{k+1} = h_{k+1}\|g_f(x_k)\|$. Since $\{g_f(x_k)\}_{k=0}^{\infty}$ is bounded, one has $\lim_{k \to \infty} \bar{h}_k = \lim_{k \to \infty} h_k = 0$.

If $\sum_{k=1}^{\infty} \bar{h}_k = +\infty$ then we may replace (2.5) with the formula $x_{k+1} = x_k - \bar{h}_{k+1} \dfrac{g_f(x_k)}{\|g_f(x_k)\|}$ and apply Theorem 2.2 to obtain the desired result.

If $\sum_{k=1}^{\infty} \bar{h}_k < \infty$ then the sequence $\{x_k\}_{k=0}^{\infty}$ converges to an certain $x^*$. We shall prove that $x^* \in M^*$. Suppose that this is not true, i.e. $0 \notin \partial f(x^*)$. Hence $\|g_f(x_k)\| \ge \varepsilon > 0$ for all sufficiently large $k$. But then $\sum_{k=1}^{\infty} \bar{h}_k$

$\geqq \varepsilon \sum_{k=1}^{\infty} h_k = +\infty$ and we obtain a contradiction. Therefore $x^* \in M^*$, which completes the proof. $\square$

Let us observe that if the function $f$ is piecewise linear (polyhedral) with a finite number of pieces, then $\{g_f(x_k)\}_{k=0}^{\infty}$ is always bounded, i.e. case (a) holds. Note also that for any $x_0$ one can always find $\delta > 0$ such that if $\max_{k \geqq 1} h_k \leqq \delta$ then case (a) must occur. Indeed, consider for some $a > 0$ the sets $U(a) = \{x : f(x) = f(x_0) + a\}$ and $S(a) = \{x : f(x) \leqq f(x_0) + a\}$. Since $f$ is defined on the whole of $E_n$ and $M^*$ is bounded, the sets $U(a)$ and $S(a)$ are compact. Consider the minimal closed sphere $S^*$ centered at $x^* \in M^*$ and containing $S(a)$, and define $c = \max_{x \in S^*} \| g_f(x) \|$, $\varrho_1 = \varrho(U(a), U(a/2))$ and $\varrho_2 = \varrho(x^*, U(a/2))$. Take $\delta = \min \{\varrho_1/c, \varrho_2/c\}$. If $h_k \leqq \delta$ then $\bar{h}_k = h_k \| g_f(x_k) \| \leqq \min \{\varrho_1, \varrho_2\}$ for all $x_k \in S^*$. We shall prove that $x_k \in S^*$ for all $k$. Indeed, $x_0 \in S^*$. If $x_k \in S(a) \backslash S(a/2)$ then (2.3) yields $\varrho(x_{k+1}, x^*) \leqq \varrho(x_k, x^*)$ and $x_{k+1} \in S^*$. If $x_k \in S(a/2)$ then $x_{k+1} \in S(a) \subseteq S^*$, since $\| x_{k+1} - x_k \| \leqq \varrho_1$. Thus we always have $x_k$ in $S^*$. But then $\| g_f(x_k) \| \leqq$ const and case (a) occurs.

The above considerations make it possible to modify Theorem 2.3 as follows.

**Theorem 2.4.** *Under the assumptions of Theorem 2.2, for any starting point $x_0$ the sequence $\{x_k\}_{k=0}^{\infty}$ generated according to the formula*

$$x_{k+1} = \begin{cases} x_k - h_{k+1}\, g_f(x_k) & \text{if } h_{k+1} \| g_f(x_k) \| \leqq c, \\ x_0 & \text{otherwise,} \end{cases}$$

*where $c$ is a positive constant, $h_k \to 0$ and $\sum_{k=1}^{\infty} h_k = +\infty$, satisfies the relations*

$$\lim_{k \to \infty} \min_{x \in M^*} \| x_k - x \| = 0; \quad \lim_{k \to \infty} f(x_k) = f^*.$$

To prove the theorem one should only observe that if $h_k \to 0$ then the condition $h_{k+1} \| g_f(x_k) \| \leqq c$ holds for all sufficiently large $k$.

**Theorem 2.5.** *Suppose that the set $M^*$ of minimum points of $f$ contains a sphere $S_r$ of radius $r > 0$ and the subgradient method (2.4) uses stepsizes $h_k > 0$ satisfying $\sum_{k=0}^{\infty} h_k = +\infty$ and $\limsup_{k \to \infty} h_k < 2r$. Then for any $x_0 \in E_n$ there exists a finite index $k(x_0)$ such that $x_{k(x_0)} \in M^*$.*

*Proof.* Let $x^* \in M^*$ be the center of the sphere $S_r$. If $x_k \notin M^*$ for $k = 0, 1, \ldots$, then $b_k(x^*) > r$ (see the proof of Theorem 2.2). By (2.3), for all $k \geqq m$ one has

$$0 \leqq \| x_{k+1} - x^* \|^2 \leqq \| x_m - x^* \|^2 - \sum_{i=m}^{k} h_{i+1}\, (2\, b_i(x^*) - h_{i+1}). \tag{2.6}$$

Since $\limsup h_k < 2\,r$, there exist $k'$ and $\delta > 0$ such that $2\,b_i(x^*) - h_{i+1} \geqq 2\,r - h_{i+1} \geqq \delta > 0$ for all $i > k'$. From (2.6) we obtain

$$0 \leqq \|x_{k'} - x^*\|^2 - \delta \sum_{i=k'}^{\infty} h_i .$$

We have arrived at a contradiction, since $\sum_{i=k'}^{\infty} h_i = +\infty$ by assumption. Therefore one can always find $x_{k(x_0)} \in M^*$, as required. $\square$

The above theorem has the following useful applications.

**Example 1.** Consider the system of convex inequalities

$$f_i(x) \leqq 0, \quad i = 1, \ldots, m, \tag{2.7}$$

where $f_i$ are convex functions defined on $E_n$.

One may solve this system by minimizing the function

$$\varphi(x) = \max \{ \max_{1 \leqq i \leqq m} f_i(x), 0 \} .$$

If the solution set $M^*$ of (2.7) has a nonempty interior, then the conditions of theorem 2.5 are fulfilled by the following iterative algorithm

$$x_{k+1} = x_k - \frac{c}{k+1} \frac{g_\varphi(x_k)}{\|g_\varphi(x_k)\|} ,$$

where $g_\varphi(x_k) = g_{f_{i*}}(x_k)$, $i^*$ is an index for which $\max_{i \in \overline{1,m}} f_i(x_k)$ is attained and $c > 0$. After a finite number of steps this algorithm will find a point $x_{k*} \in M^*$.

**Example 2** (the assignment problem [44]). The problem is to find $x_{ij} \geqq 0$, $i, j = 1, \ldots, n$, which satisfy the constraints

$$\sum_{i=1}^{n} x_{ij} = 1, \quad \sum_{j=1}^{n} x_{ij} = 1,$$

and minimize the linear form

$$\sum_{i=1}^{n} \sum_{j=1}^{n} c_{ij} x_{ij} .$$

The dual problem consists in maximizing

$$\sum_{j=1}^{n} v_j - \sum_{i=1}^{n} u_i \tag{2.8}$$

subject to

$$v_j - u_i \leqq c_{ij}, \quad i, j = 1, \ldots, n. \tag{2.9}$$

Clearly, at the optimal solution one has

$$v_j^* = \min_{1 \leqq i \leqq n} [u_i^* + c_{ij}], \tag{2.10}$$

and the dual problem resolves itself into the problem of maximizing the piecewise linear function

$$\varphi(u) = \sum_{j=1}^{n} \min_{1 \leq i \leq n} [u_i + c_{ij}] - \sum_{i=1}^{n} u_i.$$

If the assignment problem has a unique solution $x_{i(j)}^* = 1$ $(i = 1, \ldots, n)$; $x_{ij}^* = 0$ for $i \neq i(j)$, where $i(j)$ is some permutation of the set $\{1, 2, \ldots, n\}$, then by the strict complementary slackness theorem there exists a solution $u^*, v^*$ to the dual such that the minimum in (2.10) is attained only at $i = i(j)$. Then for $u$ in some neighborhood of $u^*$ one has

$$\varphi(u) = \sum_{j=1}^{n} [u_{i(j)} + c_{i(j),j}] - \sum_{i=1}^{n} u_i = \sum_{j=1}^{n} c_{i(j),j} = \varphi(u^*).$$

Consequently, if the assignment problem has a unique solution then the set of solutions of the dual problem contains an $n$-dimensional sphere. In that case Theorem 2.5 implies that the subgradient method applied to the dual problem terminates after a finite number of iterations. Thus we obtain a simple algorithm for solving the dual problem:

$$u_{k+1} = u_k + \frac{1}{k+1} g_\varphi(u_k),$$

where $g_\varphi(u_k) = \{|J_i| - 1\}_{i=1}^{n}$, $J_i = \{j: \min_k [u_k + c_{kj}] = u_i + c_{ij}\}$ and $|J_i|$ is the number of elements of $J_i$.

Unfortunately, when the assignment problem has more than one solution, the subgradient method for the dual problem will not, in general, terminate after a finite number of iterations. Therefore the above result cannot be extended to the transportation problem which generalizes the assignment problem. Indeed, although every linear transportation problem may be reduced with a given accuracy to an assignment problem, the latter one has, as a rule, many solutions.

Theorem 2.2 was generalized by B. T. Polyak [64], who considered a subgradient projection method for minimizing, subject to rather general constraints, convex functionals defined on a Hilbert space.

Let $f$ be a convex functional defined on a convex subset $Q$ of Hilbert space $E$. Assume that $Q = Q_1 \cap Q_2$, where $Q_1 = \{x: \varphi(x) \leq 0\}$, $\varphi$ is convex functional on $E$, $Q_1$ has a nonvoid interior $Q_1^{(0)}$, $Q_2$ is convex and closed, and $Q_1^{(0)} \cap Q_2 \neq \emptyset$. We denote by $P_{Q_2}(x)$ the projection of a point $x$ on $Q_2$, i.e.

$$P_{Q_2}(x) \in Q_2; \quad \|x - P_{Q_2}(x)\| = \inf_{y \in Q_2} \|x - y\|.$$

The subgradient projection method consists in generating $\{x_k\}_{k=1}^{\infty}$ by the formula

$$x_{k+1} = P_{Q_2}(x_k + \Delta(x_k)), \tag{2.11}$$

where $\Delta(x_k)$ is supporting functional of the set $\{x: f(x) \leq f(x_k)\}$ or of the set $\{x: \varphi(x) \leq \varphi(x_k)\}$, according as $x_k \in Q_1$ or $x_k \notin Q_1$. (A linear continuous functional $c$ is called a *supporting functional* of $R \subset E$ at a point $x \in R$ if $(c, x) \leqq (c, y)$ for all $y \in R$.)

**Theorem 2.6.** *Suppose that* $\lim\limits_{k \to \infty} \Delta(x_k) = 0$ *and* $\sum_{k=0}^{\infty} \| \Delta(x_k) \| = + \infty$ *in* (2.11). *Then for any* $x_0$ *one can find a subsequence* $\{x_{n_k}\}$ *such that* $x_{n_k} \in Q$ *and* $\lim\limits_{k \to \infty} f(x_{n_k}) = f^* = \inf\limits_{x \in Q} f(x)$.

*Proof.* Take any $\alpha > f^*$ and define the set $S_\alpha = \{x : x \in Q_1 ; f(x) \leqq \alpha\}$. Denote by $S_\alpha^{(0)}$ the interior of $S_\alpha$. One can always find a point $\tilde{x} \in Q_2 \cap S_\alpha^{(0)}$. Take $\varrho > 0$ such that $x \in S_\alpha$ for all $\| x - \tilde{x} \| \leqq \varrho$. Suppose that $x_k \notin S_\alpha$ for all $k$. If $x_k \notin Q_1$ then $S_\alpha \subset Q_1 \subset \{x : \varphi(x) \leqq \varphi(x_k)\}$; if $x_k \in Q_1$ then $S_\alpha \subset \{x : f(x)$ $\leqq f(x_k)\}$. In both cases $(\Delta(x_k), x_k) \leqq \left( \Delta(x_k), \tilde{x} + \dfrac{\varrho \Delta(x_k)}{\| \Delta(x_k) \|} \right)$. Hence

$$\| x_{k+1} - \tilde{x} \|^2 = \| P_{Q_2}(x_k + \Delta(x_k)) - \tilde{x} \|^2 \leqq \| x_k + \Delta(x_k) - \tilde{x} \|^2$$

$$= \| x_k - \tilde{x} \|^2 + \| \Delta(x_k) \|^2$$

$$+ 2 \left( \Delta(x_k), x_k - \tilde{x} + \varrho \frac{\Delta(x_k)}{\| \Delta(x_k) \|} \right) - 2 \varrho \| \Delta(x_k) \|$$

$$\leqq \| x_k - \tilde{x} \|^2 + \| \Delta(x_k) \|^2 - 2 \varrho \| \Delta(x_k) \| .$$

Take $N$ such that $\| \Delta(x_k) \| \leqq \varrho$ for all $k \geqq N$ and sum up the above inequalities from $k = N$ to $k = N + m$. We obtain

$$0 \leqq \| x_{N+m+1} - \tilde{x} \|^2 \leqq \| x_N - \tilde{x} \|^2 + \sum_{k=N}^{N+m} \| \Delta(x_k) \| (\| \Delta(x_k) \| - 2 \varrho)$$

$$\leqq \| x_N - \tilde{x} \|^2 - \varrho \sum_{k=N}^{N+m} \| \Delta(x_k) \| , \quad m = 1, 2, \ldots,$$

which contradicts the divergence of $\sum_{k=1}^{\infty} \| \Delta(x_k) \|$. Consequently, for every $\alpha > f^*$ one can find $x_k \in S_\alpha$, i.e. a subsequence $\{x_{n_k}\}_{k=1}^{\infty}$ such that $\lim\limits_{k \to \infty} f(x_{n_k}) = f^*$. The proof is complete. $\square$

## 2.3 On the Linear Rate of Convergence of the Subgradient Method

Under certain additional assumptions it is possible to construct versions of the subgradient method which converge at a linear rate.

**Theorem 2.7.** *Let $f$ be a convex function defined on $E_n$. Assume that for some $\varphi$ satisfying $0 \leqq \varphi < \frac{\pi}{2}$ and for all $x \in E_n$ the following inequality holds:*

$$(g_f(x), x - x^*(x)) \geqq \cos \varphi \| g_f(x) \| \| x - x^*(x) \| , \tag{2.12}$$

*where $x^*(x)$ is the point in the set of minima of $f$ that is nearest to $x$. If for a given $x_0$ we choose a stepsize $h_1$ satisfying*

$$h_1 \geq \begin{cases} \|x^*(x_0) - x_0\| \cos \varphi & \text{for } \frac{\pi}{4} \leq \varphi < \frac{\pi}{2} \qquad (2.13) \\[2mm] \dfrac{\|x^*(x_0) - x_0\|}{2 \cos \varphi} & \text{for } 0 \leq \varphi < \frac{\pi}{4}, \qquad (2.14) \end{cases}$$

*define $\{h_k\}_{k=1}^{\infty}$ by*

$$h_{k+1} = h_k \, r(\varphi), \quad k = 1, 2, \ldots, \qquad (2.15)$$

*where*

$$r(\varphi) = \begin{cases} \sin \varphi & \text{for } \frac{\pi}{4} \leq \varphi < \frac{\pi}{2}, \\[2mm] 1/2 \cos \varphi & \text{for } 0 \leq \varphi < \frac{\pi}{4}, \end{cases} \qquad (2.16)$$

*and generate $\{x_k\}_{k=0}^{\infty}$ according to the formula*

$$x_{k+1} = x_k - h_{k+1} \frac{g_f(x_k)}{\|g_f(x_k)\|},$$

*then either $g_f(x_{k^*}) = 0$ for some $k^*$. i.e. $x_{k^*}$ is a minimum point, or for all $k = 0, 1, 2, \ldots$ the following inequality holds:*

$$\|x_k - x^*(x_k)\| \leq \begin{cases} h_{k+1}/\cos \varphi & \text{for } \frac{\pi}{4} \leq \varphi < \frac{\pi}{2}, \qquad (2.18) \\[2mm] 2 h_{k+1} \cos \varphi & \text{for } 0 \leq \varphi < \frac{\pi}{4}. \qquad (2.19) \end{cases}$$

*Proof.* Let us at first consider the case $\frac{\pi}{4} \leq \varphi < \frac{\pi}{2}$. For $k = 0$ inequality (2.18) is satisfied. Suppose that it holds for $k = p$ and $g_f(x_p) \neq 0$. We shall prove it for $k = p + 1$:

$$\|x_{p+1} - x^*(x_{p+1})\|^2 \leq \|x_{p+1} - x^*(x_p)\|^2$$

$$= \left\| x_p - h_{p+1} \frac{g_f(x_p)}{\|g_f(x_p)\|} - x^*(x_p) \right\|^2$$

$$= \|x_p - x^*(x_p)\|^2 - \frac{2 h_{p+1}}{\|g_f(x_p)\|} (x_p - x^*(x_p), g_f(x_p)) + h_{p+1}^2.$$

Applying successively inequalities (2.12) and (2.18) and taking into account that $1 - 2 \cos^2 \varphi \geq 0$ for $\frac{\pi}{4} \leq \varphi < \frac{\pi}{2}$, we obtain

$$\|x_{p+1} - x^*(x_{p+1})\|^2 \leq \|x_p - x^*(x_p)\|^2 (1 - 2 \cos^2 \varphi) + h_{p+1}^2$$

$$\leq \frac{h_{p+1}^2}{\cos^2 \varphi} (1 - 2 \cos^2 \varphi) + h_{p+1}^2 = h_{p=1}^2 \left( \frac{\sin \varphi}{\cos \varphi} \right)^2.$$

In view of (2.15) and (2.16),

$$\|x_{p+1} - x^*(x_{p+1})\|^2 \leq \frac{h_{p+2}^2}{\cos^2 \varphi},$$

which yields the desired result for $\frac{\pi}{4} \leq \varphi < \frac{\pi}{2}$.

Let us now consider the case $0 \le \varphi < \frac{\pi}{4}$. For $k = 0$ (2.19) holds. Assuming that it holds for $k = p$ with $g_f(x_p) \ne 0$, we shall prove it for $k = p + 1$.

Using (2.12), (2.19) and (2.17) we obtain

$$\| x_{p+1} - x^*(x_{p+1}) \|^2 \le \| x_p - x^*(x_p) \|^2 - 2 h_p \cos \varphi \, \| x_p - x^*(x_p) \| + h_{p+1}^2$$
$$\le \| x_p - x^*(x_p) \|^2 - \| x_p - x^*(x_p) \|^2 + h_{p+1}^2 = h_{p+1}^2$$
$$= (2 \cos \varphi \, h_{p+2})^2 .$$

The proof is complete. $\square$

**Remark.** It can be seen from the proof that the theorem remains valid if (2.12) holds only for the points of the sequence $\{x_k\}_{k=0}^{\infty}$.

We conclude from the above theorem that if the angle $\varphi$ is known in advance then by choosing stepsizes according to (2.16) and (2.17) one can obtain convergence with the speed of a geometrical progression with the ratio $q = r(\varphi)$.

In formula (2.12) the value of the coefficient $\cos \varphi$ depends on how oblong the level surfaces of $f$ are. If there is no $\varphi < \frac{\pi}{2}$ such that (2.12) holds for all $x$ in a neighborhood of the optimal point of $f$, then we shall say that $f$ is *essentially gully-shaped*. In that case the stepsize rule described above does not apply, and the general rule of Theorem 2.2 should be used.

We shall now formulate a theorem analogous to Theorem 2.2 directly in terms characterizing the shape of the level surfaces.

**Theorem 2.8.** *Let $f$ be a convex function defined on $E_n$ and let $x^*$ be its unique minimum point. Assume that for a given starting point $x_0$ numbers $\sigma$ and $h_1$ are chosen such that $\sigma \ge \sqrt{2}$, $h_1 \ge \| x_0 - x^* \| / \sigma$. Furthermore, let $Y = \{ y : \| y - x^* \| \le \sigma h_1 \}$ and let for every pair of points $x, z \in Y$ satisfying $f(x) = f(z) \ne f(x^*)$ the following inequality hold:*

$$\| x - x^* \| \le \sigma \| z - x^* \| . \tag{2.20}$$

*Then the sequence $\{x_k\}_{k=0}^{\infty}$ generated recursively by*

$$x_{k+1} = x_k - h_{k+1} \frac{g_f(x_k)}{\| g_f(x_k) \|} ,$$

$$h_{k+1} = h_k \frac{\sqrt{\sigma^2 - 1}}{\sigma} , \qquad k = 0, 1, 2, \ldots ,$$

*converges to $x^*$ at a linear rate:*

$$\| x_k - x^* \| \le h_{k+1} \sigma , \tag{2.21}$$

*except for the case in which $g_f(x_{\bar{k}}) = 0$ for some $\bar{k}$, i.e. $x_{\bar{k}} = x^*$.*

*Proof.* For $k = 0$ inequality (2.21) holds, since $x_0 \in Y$. Assuming that (2.21) is valid for $k = p$, we shall prove it for $k = p + 1$. If $x_p \ne x^*$ then (2.20) holds, because $x_p \in Y$. Next,

$$(g_f(x_p), x_p - x^*) = a_p(x^*) \| g_f(x_p) \| ,$$

where $a_p(x^*)$ is the projection of $x_p - x^*$ on $g_f(x_p)$. From the proof of Theorem 2.1 we know that $a_p(x^*) \geq b_p(x^*)$, where $b_p(x^*)$ is the distance from $x^*$ to the level surface $f(x) = f(x_p)$. Thus $(g_f(x_p), x_p - x^*) \geq b_p(x^*) \|g_f(x_p)\|$, and (2.20) yields $b_p(x^*) \geq \|x_p - x^*\|/\sigma$. Hence

$$(g_f(x_p), x_p - x^*) \geq \frac{1}{\sigma} \|g_f(x_p)\| \ \|x_p - x^*\|. \tag{2.22}$$

Set $\cos \varphi = 1/\sigma$. Then condition (2.22) assumes the form (2.12). Furthermore, $\varphi \geq \frac{\pi}{4}$ and $\sin \varphi = \sqrt{\sigma^2 - 1}/\sigma$ for $\sigma \geq \sqrt{2}$, so one can obtain $\|x_{p+1} - x^*\| \leq h_{p+2} \sigma$ as in the proof of Theorem 2.7. Consequently, $\|x_k - x^*\| \leq h_{k+1} \sigma$ for all $k$, as required. $\square$

**Example.** Consider the special case of minimizing a positive definite quadratic form $f(x) = \frac{1}{2}(A x, x)$. Denote by $\lambda$ and $\mu$ the smallest and the largest eigenvalues of the matrix $A$, respectively. Let us calculate the minimum value of the quotient

$$\frac{(g_f(x), x)}{\|g_f(x)\| \ \|x\|} = \frac{(A x, x)}{\|A x\| \ \|x\|}, \quad \|x\| \neq 0.$$

To this end, choose any vector $x_0$ different from all the eigenvectors of $A$ and consider the two-dimensional subspace $E(x_0)$ spanned by $x_0$ and $A x_0$. Let us denote by $\tilde{A}$ the matrix of the quadratic form defined on $E(x_0)$ that satisfies $(\tilde{A}x, x) = (A x, x)$ for all $x \in E(x_0)$. It is easy to observe that the operator $\tilde{A}$ defined on the subspace $E(x_0)$ is positive definite and its eigenvalues $\tilde{\lambda}$ and $\tilde{\mu}$ ($\tilde{\lambda} \leq \tilde{\mu}$) satisfy the inequalities $\lambda \leq \tilde{\lambda} \leq \tilde{\mu} \leq \mu$. Let $x_1$ and $x_2$ be the orthonormal eigenvectors of $A$ in $E(x_0)$ and assume that $\|x_0\| = 1$. Let us express $x_0$ as $x_0 = \alpha_1 x_1 + \alpha_2 x_2$, where $\alpha_1^2 + \alpha_2^2 = 1$. Then

$$\frac{(A x_0, x_0)}{\|A x_0\| \ \|x_0\|} = \frac{(\tilde{A} x_0, x_0)}{\|\tilde{A} x_0\| \ \|x_0\|} = \frac{\alpha_1^2 \tilde{\lambda} + \alpha_2^2 \tilde{\mu}}{\sqrt{\alpha_1^2 \tilde{\lambda}^2 + \alpha_2^2 \tilde{\mu}^2}}. \tag{2.23}$$

Minimizing the right side of (2.23) subject to the constraint $\alpha_1^2 + \alpha_2^2 = 1$ we obtain the inequality

$$\frac{(A x_0, x_0)}{\|A x_0\| \ \|x_0\|} \geq \frac{2\sqrt{\tilde{\lambda} \tilde{\mu}}}{\tilde{\lambda} + \tilde{\mu}},$$

with equality occuring for $\alpha_1^2 = \tilde{\mu}/(\tilde{\lambda} + \tilde{\mu})$, $\alpha_2^2 = \tilde{\lambda}/(\tilde{\lambda} + \tilde{\mu})$. This immediately implies that

$$\min_{x \in E_n, x \neq 0} \frac{(A x, x)}{\|A x\| \ \|x\|} = \frac{2\sqrt{\lambda \mu}}{\lambda + \mu},$$

with the minimum attained at

$$x = \sqrt{\frac{\mu}{\lambda + \mu}} \, s_1 + \sqrt{\frac{\lambda}{\lambda + \mu}} \, s_2,$$

where $s_1$ and $s_2$ are the orthonormal eigenvectors of $A$ that correspond to the smallest and the largest eigenvalues, respectively.

Let us now return to Theorem 2.7. In our example we may choose $\cos \varphi = 2 \sqrt{\lambda \mu}/(\lambda + \mu)$. Then $\sin \varphi = (\mu - \lambda)/(\mu + \lambda)$ and the subgradient method described in that theorem generates a sequence converging to the optimal point at a linear rate with the ratio (for $\varphi \geq \frac{\pi}{4}$)

$$q = \frac{\mu - \lambda}{\mu + \lambda} = \frac{\varrho - 1}{\varrho + 1},$$

where $\varrho = \mu/\lambda$ is the *condition number* of $A$ [29]. Additionally the sequence of the function values converges linearly with the ratio $q^2$.

**Remark.** Almost all our considerations remain valid if $A$ is a nonnegative definite operator, $A \neq 0$; the only difference is that $\lambda$ should be defined as the smallest positive eigenvalue.

Let us compare the result obtained above with the result of L. V. Kantorovich [45], which concerns the rate of convergence of the method of steepest descent for quadratic functionals. In that method the functional values also converge linearly with the ratio $q^2$. However, the method of steepest descent for minimizing quadratic functionals (or for solving systems of linear equations with a symmetric, positive definite matrix) is far more complicated than the method described in our theorem: it requires $3n$ storage cells, i.e. three times more than our method, which is significant for large scale problems (see [29, § 70]).

Therefore if we can obtain a sufficiently tight upper bound $\hat{\varrho}$ of the condition number $\varrho$ and if we take $\sin \varphi = (\hat{\varrho} - 1)/(\hat{\varrho} + 1)$, then the algorithm of Theorem 2.7 may compete in practice with the method of steepest descent.

Let us note that for a quadratic function $f(x) = (Ax, x)$ the number $\sigma$ involved in Theorem 2.8 may be taken equal to $\sqrt{\varrho}$, ensuring linear convergence with the ratio

$$\bar{q} = \frac{\sqrt{\sigma^2 - 1}}{\sigma} = \sqrt{\frac{\varrho - 1}{\varrho}}.$$

We shall now consider another version of the subgradient method, in which the stepsize remains constant for a certain number of iterations and then is halved [94].

**Theorem 2.9.** *Let for a convex function f the assumptions of Theorem 2.8 be fulfilled with* $\sigma \geq 2$. *For a given* $x_0$ *consider the iterative algorithm:*

$$x_{k+1} = x_k - h_{k+1} \frac{g_f(x_k)}{\| g_f(x_k) \|},$$

*where* $h_{k+1} = h_0 \, 2^{-[(k+1)/N]}$. *If* $h_0$ *is large enough and* $N \geq 3\sigma^2 + 1$ *then*

$$\| x_k - x^* \| \leq 2\sigma h_{k+1}, \quad k = 0, 1, 2, \ldots . \tag{2.24}$$

*Proof.* Let $h_0 \geq \|x_0 - x^*\|/2\,\sigma$. Just as in the proof of Theorem 2.1, we obtain

$$\|x_{k+1} - x^*\|^2 \leq \|x_k - x^*\|^2 + h_{k+1}^2 - 2\,h_{k+1}\,b_k(x^*).$$

If $b_k(x^*) \geq h_{k+1}/2$ then

$$\|x_{k+1} - x^*\| \leq \|x_k - x^*\|, \tag{2.25}$$

while for $b_k(x^*) \geq h_{k+1}$ we get

$$\|x_{k+1} - x^*\|^2 \leq \|x_k - x^*\|^2 - h_{k+1}^2. \tag{2.26}$$

Set $h_i = h_0$, $i = 1, \ldots, [3\,\sigma^2] + 1$. Suppose that (2.26) holds for all iterations. Since $\|x_0 - x^*\|^2 \leq 4\,\sigma^2\,h_0^2$, we obtain

$$\|x_N - x^*\|^2 \leq 4\,\sigma^2\,h_0^2 - ([3\,\sigma^2] + 1)\,h_0^2 \leq \sigma^2\,h_0^2.$$

If inequality (2.26) is violated for a certain $\bar{k} < N$, then $b_{\bar{k}}(x^*) \leq h_0$ and thus $\|x_{\bar{k}} - x^*\| \leq \sigma\,h_0$. Then two cases are possible.

(a) The inequality $b_k(x^*) \geq h_{k+1}/2$ holds for $k = \bar{k} + 1, \ldots, N$. In this case (2.25) yields $\|x_N - x^*\| \leq \sigma\,h_0$.

(b) At a certain step $\bar{\bar{k}}, \bar{k} \leq \bar{\bar{k}} \leq N$, one has $b_{\bar{\bar{k}}}(x^*) < h_0/2$. Then $\|x_{\bar{\bar{k}}} - x^*\| \leq h_0\,\sigma/2$ and $\|x_k - x^*\| \leq (h_0\,\sigma/2) + h_0 \leq h_0\,\sigma$ for $\bar{\bar{k}} \leq k \leq N$ (recall that $\sigma \geq 2$).

Therefore, in all cases

$$\|x_N - x^*\| \leq \sigma\,h_0; \quad \|x_i - x^*\| \leq 2\,\sigma\,h_0, \quad i = 0, \ldots, N.$$

Setting $x_0^{(1)} = x_N$ and $h_0^{(1)} = h_0/2$, we obtain the inequality $h_0^{(1)} \geq \|x_0^{(1)} - x^*\|/2\,\sigma$, analogous to (2.24), which implies that $\|x_i - x^*\| \leq \sigma\,h_0$, $i = N + 1, \ldots, 2\,N - 1$;

$$\|x_{2N} - x^*\| \leq \frac{\sigma\,h_0}{2}.$$

By induction,

$$\|x_{rN+i} - x^*\| \leq 2\,\sigma\,h_0\,2^{-r}, \quad i = 0, \ldots, N - 1, \quad r = 1, 2, \ldots,$$

as required.   $\square$

Let us address the question of selecting the parameter values of the version of the subgradient method described in Theorem 2.7. For simplicity we shall assume that $f$ attains its minimum at a unique point $x^*$. The algorithm requires values of the two parameters $h$ and $\varphi$. If the value of $\varphi$ is chosen so that inequality (2.12) holds for all $x = x_k$, $k = 0, 1, \ldots$, and $h$ satisfies inequality (2.13), then Theorem 2.7 ensures that the following accuracy is attained after $k$ steps:

$$\Delta_k = \|x_k - x^*\| \leq \frac{h\sin^k\varphi}{\cos\varphi}. \tag{2.27}$$

However, in most cases it is difficult to verify condition (2.12), and so one needs a certain indirect, easily verifiable criterion of the "correctness" of the choice of $h$ and $\varphi$.

Take another starting point $\bar{x}_0$ and $\bar{\varphi}$, $\varphi \leq \bar{\varphi} < \frac{\pi}{2}$. Set $\bar{h} = h + \|x_0 - \bar{x}_0\| \cos \varphi$. It is easy to observe that if the parameters $\varphi$ and $h$ satisfy (2.12) and (2.13) for the point $x_0$, then those conditions hold also for $\bar{\varphi}, \bar{h}$. Let $\{\bar{x}_k\}$, $k = 0, 1, \ldots$, be a sequence generated by the subgradient method from the starting point $\bar{x}_0$ with the parameters $\bar{\varphi}$ and $\bar{h}$. Then (2.27) yields

$$\|x_k - \bar{x}_k\| \leq \frac{h \sin^k \varphi}{\cos \varphi} + \frac{h \sin^k \bar{\varphi}}{\cos \bar{\varphi}}, \quad k = 0, 1, \ldots . \tag{2.28}$$

If the above inequality is violated for some $k$, this indicates that the values chosen for $h$ and $\varphi$ are wrong. On the other hand, if at least one of the sequences $\{x_k\}$, $\{\bar{x}_k\}$ does not converge to $x^*$, then, in view of $\bar{x}_0 \neq x_0$ and $\bar{\varphi} \neq \varphi$, it is highly improbable that these sequences have equal limits, while if the limits differ then inequality (2.28) must be violated for a sufficiently large $k$.

Therefore, inequality (2.28) can be used for verifying the correctness of the choice of the parameter values, provided that two trajectories are generated parallelly by the subgradient method from two different starting points. One can apply this test as follows. Suppose that we seek $x_{\bar{k}}$ such that

$$\|x_{\bar{k}} - x^*\| \leq \delta. \tag{2.29}$$

By (2.27), if $h$ and $\varphi$ are well chosen then to ensure (2.29) it is sufficient that

$$\bar{k} > \frac{\ln(\delta \cos \varphi / h)}{\ln(\sin \varphi)}. \tag{2.30}$$

Let us start the algorithm from two points $x_0$ and $\bar{x}_0$, as described above. If inequality (2.28) holds for $\bar{k}$ consecutive iterations, where $\bar{k}$ satisfies (2.30), then one may claim that $\|x_{\bar{k}} - x^*\| \leq \delta$. For security one may start the algorithm from yet another point, or explore the neighborhood of the point $x_{\bar{k}}$. On the other hand, if for some $k \leq \bar{k}$ inequality (2.28) is violated, then one should change the parameters $h$ and $\varphi$ (e.g. increase $\varphi$ so as to halve $\cos \varphi$) and continue the computation from the best point obtained so far and from a certain trial point.

## 2.4 The Subgradient Method and Fejer-type Approximations

We shall establish useful relations between the subgradient method and the method of Fejer-type approximations developed by I. I. Eremin.

**Definition** [23]. *A mapping $\varphi: E_n \to E_n$ is called M-Fejer ($E_n \supset M \neq \emptyset$), if*

$$\varphi(y) = y \quad and \quad \|\varphi(x) - y\| < \|x - y\| \quad for\ all\ y \in M\ and\ x \notin M. \tag{2.31}$$

Using $M$-Fejer maps, one can construct Fejer-type algorithms of the form $x_{k+1} = \varphi(x_k)$, which converge to $M$.

**Theorem 2.10** [23]. *If an $M$-Fejer map $\varphi$ is continuous then for any starting point $x_0$ the sequence $\{x_k\}$ generated by $x_{k+1} = \varphi(x_k)$, $k = 0, 1, \ldots$, converges to an element of $M$.*

*Proof.* Fix $y \in M$. By (2.31), $\|x_k - y\| \le \|x_0 - y\|$, i.e. the sequence $\{x_k\}_{k=0}^{\infty}$ is bounded. Let $x^*$ be an accumulation point of this sequence. Suppose that $x^* \notin M$. Then $\|\varphi(x^*) - y\| < \|x^* - y\|$ and it follows from the continuity of $\varphi$ that one can find a sufficiently large $\bar{k}$ for which $\|\varphi(x_{\bar{k}-1}) - y\| = \|x_{\bar{k}} - y\| < \|x^* - y\|$. On the other hand, $\|x^* - y\|$ is the limit of a decreasing sequence $\{\|x_k - y\|\}$, i.e. $\|x^* - y\| \le \|x_k - y\|$ for all $k=0,1,\ldots$. We have obtained a contradiction. Thus $x^* \in M$ and the proof is complete. $\square$

B. T. Polyak [65] suggested versions of the subgradient method which are essentially Fejer-type algorithms.

Let $f$ be a convex function defined on $E_n$, and let $f^* = \inf f(x)$, $c \ge f^*$, $M(c) = \{x \in E_n : f(x) \le c\}$. Consider the following iterative algorithm:

$$x_{k+1} = x_k - \frac{\gamma[f(x_k) - c]}{\|g_f(x_k)\|^2} g_f(x_k) = \varphi_c(x_k). \tag{2.32}$$

**Theorem 9.11.** *If $0 < \gamma < 2$ then for any $x_0 \in E_n$ either $x_{k*} \in M(c)$ for some $k^*$ or $\lim\limits_{k \to \infty} x_k \in M(c)$.*

*Proof.* Let $x_k \notin M(c)$ and $y \in M(c)$. Then

$$\|x_{k+1} - y\|^2 = \|x_k - y\|^2 - 2\gamma \frac{f(x_k) - c}{\|g_f(x_k)\|^2} (x_k - y, g_f(x_k))$$

$$+ \gamma^2 \frac{[f(x_k) - c]^2}{\|g_f(x_k)\|^2}.$$

Observe that

$$0 \le f(x_k) - c \le f(x_k) - f(y) \le (g_f(x_k), x_k - y),$$

which yields

$$\|x_{k+1} - y\|^2 \le \|x_k - y\|^2 - \gamma(2 - \gamma) \frac{[f(x_k) - c]^2}{\|g_f(x_k)\|^2} < \|x_k - y\|^2. \tag{2.33}$$

Setting $\varphi_c(y) = y$ for all $y \in M(c)$, we conclude that $\varphi_c(x)$ is an $M(c)$-Fejer map. Actually, Theorem 2.10 is not directly applicable here, because the map $\varphi_c(x)$ is not necessarily continuous (the subgradient mapping $g_f$ may be discontinuous).

Let us observe, however, that the monotonicity of $\{\|x_k - y\|\}$ implies the boundedness of $\{x_k\}_{k=0}^{\infty}$ and $\{g_f(x_k)\}_{k=0}^{\infty}$. If an accumulation point $x^*$ of $\{x_k\}_{k=0}^{\infty}$ satisfies $f(x^*) > c$, then there exist $\delta > 0$ and an infinite sequence of indices $k_1 < k_2 < \ldots$ such that $f(x_{k_i}) - c > \delta$, i.e.

$$\| x_{k_i+1} - y \|^2 \le \| x_{k_i} - y \|^2 - \varepsilon, \quad i = 1, 2, \ldots,$$

where $\varepsilon > 0$. Hence $\| x_{k_i+1} - y \|^2 \le \| x_{k_0} - y \|^2 - i\,\varepsilon$. We have arrived at a contradiction. Therefore, $x^* \in M(c)$. The uniqueness of $x^*$ results from the monotone convergence of the distance $\| x_k - x^* \|$. The proof is complete. $\square$

Under rather restrictive assumptions the method (3.23) converges with the speed of a geometrical progression.

**Theorem 2.12.** *Let $f$ be a strongly convex function satisfying $f(x) - f^* \ge m \| x - x^* \|^2$ for some $m > 0$ and all $x$, and let its gradient $g_f(x)$ be Lipschitz continuous with a constant $L$ in the area $\| x - x^* \| \le \| x_0 - x^* \|$. Then for the method (2.32) with $c = f^*$ one has*

$$\| x_k - x^* \| \le q^k \| x_0 - x^* \|,$$

*where*

$$q = \left( 1 - \gamma(2 - \gamma) \frac{m^2}{L^2} \right)^{1/2} < 1.$$

*Proof.* By the assumptions,

$$f(x_k) - f(x^*) \ge m \| x_k - x^* \|^2, \tag{2.34}$$

$$\| g_f(x^k) \| \le L \| x_k - x^* \|. \tag{2.35}$$

From (2.33)–(2.35) we obtain

$$\| x_{k+1} - x^* \|^2 \le \| x_k - x^* \|^2 - \frac{\gamma(2 - \gamma)\, m^2}{L^2} \frac{\| x_k - x^* \|^4}{\| x_k - x^* \|^2}$$

$$= \| x_k - x^* \|^2 \left[ 1 - \gamma(2 - \gamma) \frac{m^2}{L^2} \right],$$

which completes the proof. $\square$

In [65] another condition guaranteeing linear convergence of (2.32) was considered.

**Theorem 2.13.** *Let $f$ be a convex function with a minimum point $x^*$, let $m > 0$ be such that $f(x) - f(x^*) \ge m \| x - x^* \|$ for all $x$, and let $L$ be the Lipschitz constant of $f$ in the area $\| x - x^* \| \le \| x_0 - x^* \|$. Then for the algorithm (2.32) with $c = f(x^*)$ one has*

$$\| x_{k+1} - x^* \| \le q \| x_k - x^* \|, \quad q = \left( 1 - \gamma(2 - \gamma) \frac{m^2}{L^2} \right)^{1/2}.$$

*Proof.* From (2.33) and the assumptions we have

$$\|x_{k+1} - x^*\|^2 \leq \|x_k - x^*\|^2 - \frac{\gamma(2-\gamma)\,m^2}{L^2} \|x_k - x^*\|^2 = q\,\|x_k - x^*\|^2,$$

whence the assertion follows. $\square$

Let us consider an application of algorithms of the form (2.32) to the solution of a consistent system of convex inequalities

$$f_i(x) \leq 0, \quad i = 1, \ldots, m. \tag{2.36}$$

The solution of this system is equivalent to minimizing the function

$$\psi^+(x) = \begin{cases} \psi(x) & \text{if } \psi(x) > 0, \\ 0 & \text{if } \psi(x) \leq 0, \end{cases}$$

where

$$\psi(x) = \max f_i(x).$$

For this problem one may take $c = 0$ in (3.32), obtaining the following algorithm

$$x_{k+1} = \begin{cases} x_k & \text{if } \psi^+(x_k) = 0, \\ x_k - \dfrac{g_{f_{i_k}}(x_k)\,\psi^*(x_k)}{\|g_{f_{i_k}}(x_k)\|^2} & \text{if } \psi^+(x_k) = f_{i_k}(x_k) > 0, \end{cases} \tag{2.37}$$

which converges to a solution from any starting point $x_0$.

Let us consider the case of an inconsistent system (2.32), i.e., $\min \psi^+(x) = d > 0$. To this end, observe that the algorithm (2.32) converges also with $\gamma$ dependent on $k$:

$$x_{k+1} = x_k - \frac{\gamma_k\,[f(x_k) - c]}{\|g_f(x_k)\|^2}\,g_f(x_k),$$

provided that $\varepsilon \leq \gamma_k \leq 2 - \varepsilon$ for all $k$, where $\varepsilon > 0$.

Applying the algorithm (2.32) with $0 < \gamma < 2$ and $c = 0$ to the function $\psi$, we obtain

$$x_{k+1} = x_k - \frac{\gamma\,\psi(x_k)}{\|g_\psi(x_k)\|^2}\,g_\psi(x_k)$$

$$= x_k - \frac{\gamma\,\psi(x_k)}{\psi(x_k) - d}\,[\psi(x_k) - d]\,\frac{g_\psi(x_k)}{\|g_\psi(x_k)\|^2}.$$

If the inequality $\gamma\,\dfrac{\psi(x_k)}{\psi(x_k) - d} \leq 2 - \varepsilon$ were satisfied for all $k$, the algorithm would converge to the minimum of $\psi$. But this is impossible because then

$$\gamma \frac{\psi(x_k)}{\psi(x_k) - d} \to \infty.$$ So, one can find $k^*$ such that

$$\gamma \frac{\psi(x_{k^*})}{\psi(x_{k^*}) - d} > 2 - \varepsilon,$$

$$\psi(x_{k^*}) < \frac{d(2 - \varepsilon)}{2 - \varepsilon - \gamma}.$$

Since $\varepsilon$ may be arbitrarily small we have the following theorem.

**Theorem 2.14.** *Let $\psi$ be a convex function defined on $E_n$, and let $\min_{x \in E_n} \psi(x) = d > 0$. Consider the sequence $\{x_k\}$ generated by the algorithm*

$$x_{k+1} = x_k - \frac{\gamma \, \psi(x_k)}{\|g_\psi(x_k)\|^2} g_\psi(x_k)$$

*from a starting point $x_0$. Then*

$$\lim_{k \to \infty} \min_{0 \le i \le k} \psi(x_i) = \frac{2 \, d}{2 - \gamma}.$$

Theorems 2.11 and 2.14 indicate that it is theoretically possible to construct an algorithm for finding a minimum of a convex function $f$, even if $f^* = \min_{x \in E_n} f(x)$ is unknown. Such an algorithm could use an estimate $\bar{f}$ of $f^*$, improving it successively on the basis of the function values $\{f(x_k)\}$ observed, with

$$x_{k+1} = x_k - \frac{\gamma [f(x_k) - \bar{f}]}{\|g_f(x_k)\|^2} g_f(x_k), \quad 0 < \gamma < 2.$$

If $f(x_k) - \bar{f} \geq d > 0$ for $k = 0, 1, 2, \ldots$, then $\bar{f} < f^*$ and the estimate $\bar{f}$ should be increased. But this test cannot be applied in practice, because there is no algorithm which could detect from a finite part of a sequence whether this sequence is convergent. One can suggest here only some heuristic rules having limited applicability.

A group of Italian scientists [13] proposed a modification of the subgradient method, which uses both an estimate of the minimum value of the function and the information about the direction applied at the previous step. Although their paper deals with the minimization of a piecewise-linear convex function, the algorithm suggested may be applied with no essential changes to any convex function.

The main idea of the method consists in computing the current search direction as a linear combination of the current subgradient and the direction used at the previous step (as in the conjugate gradient method).

Let $f$ be a convex function on $E_n$, and let $f^* = \min\limits_{x \in E_n} f(x) = f(x^*)$. Consider the following procedure:

$$x_{k+1} = x_k - h_k s_k; \quad h_k = \frac{[f(x_k) - f^*] \gamma_k}{\| s_k \|^2}; \tag{2.38}$$

$$s_0 = g_f(x_0); \quad s_k = g_f(x_k) + \beta_k s_{k-1}; \quad k = 1, 2, \ldots.$$

**Theorem 2.15.** *If $0 < \gamma_k \le 1$ and $\beta_k \ge 0$ for all $k = 0, 1, \ldots,$ then $(x_k - x^*, s_k) \ge (x_k - x^*, g_f(x_k))$ for all $k$.*

*Proof.* We shall prove the theorem by induction. Suppose that the assertion of the theorem is true for some $k = m$. To prove it for $k = m + 1$ observe that

$$(x_{m+1} - x^*, s_{m+1}) = (x_m - h_m s_m - x^*, g_f(x_{m+1}) + \beta_m s_m)$$

$$\ge (x_{m+1} - x^*, g_f(x_{m+1})) + \beta_m [(x_m - x^*, s_m) - (f(x_m) - f^*)].$$

By the inductive assumptions

$$(x_m - x^*, s_m) \ge (x_m - x^*, g_f(x_m)) \ge f(x_m) - f^*.$$

Since $\beta_m \ge 0$, the two preceding inequalities yield

$$(x_{m+1} - x^*, s_{m+1}) \ge (x_{m+1} - x^*, g_f(x_{m+1})),$$

which completes the proof. $\square$

With a particular choice of $\beta_k$ one can obtain a stronger result.

**Theorem 2.16.** *Under the conditions of the preceding theorem, let*

$$\beta_k = \begin{cases} -\alpha_k \dfrac{(s_{k-1}, g_f(x_k))}{\| s_{k-1} \|^2} & \text{if } (s_{k-1}, g_f(x_k)) < 0; \\ 0 & \text{otherwise,} \end{cases}$$

$$0 \le \alpha_k \le 2.$$

*Then*

$$\frac{(x_k - x^*, s_k)}{\| s_k \|} \ge \frac{(x_k - x^*, g_f(x_k))}{\| g_f(x_k) \|}. \tag{2.39}$$

*Proof.* If $\beta_k = 0$ then $s_k = g_f(x_k)$ and (2.39) holds. Consider the case

$$(s_{k-1}, g_f(x_k)) < 0,$$

$$\beta_k = \frac{-\alpha_k (s_{k-1}, g_f(x_k))}{\| s_{k-1} \|^2}.$$

Then

$$\| s_k \|^2 - \| g_f(x_k) \|^2 = \beta_k^2 \| s_{k-1} \|^2 + 2 \beta_k (s_{k-1}, g_f(x_k))$$

$$= \frac{(s_{k-1}, g_f(x_k))^2}{\| s_{k-1} \|^2} (\alpha_k^2 - 2 \alpha_k) \le 0.$$

Hence $\| s_k \| \leqq \| g_f(x_k) \|$. Applying Theorem 2.15, we obtain

$$\frac{(s_k, x_k - x^*)}{\| s_k \|} \geqq \frac{(g_f(x_k), x_k - x^*)}{\| g_f(x_k) \|},$$

as required.   □

We see from the proof of the theorem that, under certain conditions, the direction $- s_k$ computed by (2.38) forms a smaller angle with the direction towards the minimum than does the direction negative to the subgradient, thus enhancing the speed of convergence for a given $f^*$. In [13] the use of $\alpha_k \sim 1.5$ is recommended and computational results are given, which indicate that the algorithm (2.38) is superior to the subgradient method (2.32).


## 2.5 Methods of $\varepsilon$-subgradients


In recent years several papers appeared in which various descent algorithms of nonsmooth optimization were proposed. Widely known are the works of C. Lemaréchal and P. Wolfe on the so-called *method of conjugate subgradients* for minimizing convex functions [52, 101]. Although that method formally resembles the conjugate gradient method, it is actually in no way connected with quadratic approximations, but in a certain sense generalizes the iterative algorithms of V. F. Demyanov, elaborated for the minimization of max functions [18]. In the algorithm of conjugate subgradients an appropriate search direction is found on the basis of the information provided by the subgradients calculated at points in a certain neighborhood of the current point. Therefore that method may be classified as an $\varepsilon$-subgradient method. Our description of that method follows [51].

Let $f$ be a convex function on $E_n$ and let $x_0 \in E_n$. Take a subgradient $g(x_0) \in G_f(x_0)$ and the direction $s_0 = - g(x_0)$, which is not necessarily a direction of descent. If $f(x_0 + \varrho s_0) \geqq f(x_0)$ for all $\varrho > 0$ then $(g(x_0 + \varrho s_0), s_0) \geqq 0$ for $\varrho > 0$.

Let $g_1$ be an accumulation point of the sequence $g(x_0 + \varrho_i s_0)$, where $\varrho_i \to 0+$. Then $(g_1, s_0) \geqq 0$ and $\| g_0 - g_1 \|^2 \geqq \| g_0 \|^2 + \| g_1 \|^2$. If $\| g_0 - g_1 \|$ is small, then so are $\| g_0 \|$ and $\| g_1 \|$, i.e., the necessary condition of optimally is nearly fulfilled. Otherwise the vector $g_1 \in G_f(x_0)$ is significantly different from $g_0$. Next, suppose that we have $g_0, \ldots, g_k \in G_f(x_0)$ and let $G_k(x_0)$ be the convex hull of $g_0, \ldots, g_k$. The set $G_k(x_0)$ may be regarded as an approximation of $G_f(x_0)$. If $G_k(x_0) = G_f(x_0)$, the direction of steepest descent at $x_0$ could be calculated as the solution to the following problem

$$\min_{\| s \| = 1} \max_{i \in \overline{0,k}} (s, g_i). \tag{2.40}$$

The problem (2.40) is equivalent to the quadratic programming problem

$$\text{minimize } \|s\|^2$$
$$\text{subject to } -s = \sum_{i=0}^{k} \lambda_i\, g_i,$$
$$\lambda_i \geqq 0, \quad \sum_{i=0}^{k} \lambda_i = 1,$$

i.e., to the problem of finding the point in $-G_k(x_0)$ that is nearest to the origin (see Theorem 1.11).

Let $s_k$ be the solution of (2.40). Then either $s_k$ is a direction of descent or one can find $g_{k+1} \in G_f(x_0)$ such that $(g_{k+1}, s_k) \geqq 0$. Continuing this process, we shall either obtain a descent direction, or construct a sequence $\{g_k\}$ converging to the origin, which will indicate the optimality of $x_0$. One could easily imagine an algorithm in which directions of descent are constructed as described above, and the algorithm descends by successive directional minimizations. However, such an algorithm would not work, because it assumes exact line searches and, being an analogue of the method of steepest descent, it can converge to a nonoptimal point.

To eliminate the above drawbacks it is necessary to replace the subdifferential $G_f(x_0)$ by the *$\varepsilon$-subdifferential* $\partial_\varepsilon f(x_0)$, defined as follows: $\partial_\varepsilon f(x_0) = \{g \in E_n : f(x) - f(x_0) \geqq (g, x - x_0) - \varepsilon \text{ for all } x \in E_n\}$.

The $\varepsilon$-subdifferential $\partial_\varepsilon f(x)$ has the following properties:

1) for any $x \in E_n$ and $\varepsilon \geqq 0$ it is nonempty, convex and compact;
2) $\partial_0 f(x) = G_f(x)$;
3) for each $\varepsilon > 0$ there exists $\delta > 0$ such that if $\|y - x\| \leqq \delta$ and $g \in G_f(y)$ then $g \in \partial_\varepsilon f(x)$ [87].

Moreover, we have the following relation which generalizes the formula for the directional derivative

$$\max_{g \in \partial_\varepsilon f(x)} (s, g) = \inf_{\varrho \geqq 0} \left\{ \frac{f(x + \varrho s) - f(x) + \varepsilon}{\varrho} \right\}.$$

Hence, if a direction $s$ exists such that

$$\max_{g \in \partial_\varepsilon f(x)} (s, g) < 0, \tag{2.41}$$

then one can find $\varrho > 0$ for which $f(x + \varrho s) < f(x) - \varepsilon$; if such a direction does not exist, then $f(x) \leqq f^* + \varepsilon$. Therefore, one may think of an algorithm for finding $\min f(x)$ with accuracy $\varepsilon$ which would compute search directions $s$ satisfying (2.41). The only difficulty is that the set $\partial_\varepsilon f(x)$ is not defined explicitly. This difficulty is overcome in Lemaréchal's algorithm [51] by approximating $\partial_\varepsilon f(x)$ with a convex hull of subgradients collected at previously generated points belonging to a certain neighborhood of $x$.

Let us pass on to the description of Lemaréchal's algorithm (we shall call it *the L-algorithm*). It is a two-level algorithm with the outer loop generating a minimizing sequence (indexed by $p$), and the inner one determining search directions (indexed by $k$).

Let $\varepsilon > 0$, $\eta > 0$, $x_0 \in E_n$, $g_0 \in \partial_0 f(x_0) = G_f(x_0)$, $p = 0$.

1. Set $s_0 = -g_0$, $k = 0$.
2. Find $\varrho_k = \arg \min_{\varrho \geqq 0} f(x_p + \varrho s_k)$ and $g_{k+1} \in G_f(x_p + \varrho_k s_k)$ such that $(g_{k+1}, s_k) = 0$.
3. If $f(x_p + \varrho_k s_k) < f(x_p) - \varepsilon$ then set $x_{p+1} = x_p + \varrho_k s_k$, $g_0 = g_{k+1}$, increase $p$ by 1 and to to Step 1; otherwise go to Step 4.
4. Compute $s_{k+1}$ by projecting the null vector onto the convex hull of $\{-g_0, \ldots, -g_{k+1}\}$.
5. If $\|s_{k+1}\| \leq \eta$ then stop; otherwise go to Step 6.
6. Increase $k$ by 1 and go to Step 2.

It is easy to prove the following theorem.

**Theorem 2.17.** *Suppose* $\lim_{\|x\| \to \infty} f(x) = +\infty$ *and let* $x(\eta)$ *denote the point found by the L-algorithm. If the sequence* $\{\eta_r\}_{r=0}^\infty$ *converges to zero, then* $\lim_{r \to \infty} f(x(\eta_r)) \leq \inf f(x) + \varepsilon$. *(It is assumed here that after completing the computation with* $\eta = \eta_r$ *the L-algorithm is run with* $\eta = \eta_{r+1}$ *from the starting point* $x(\eta_r)$.)

*Proof.* Because $\lim_{\|x\| \to \infty} f(x) = +\infty$ and the L-algorithm is a descent method, the sequence $\{x(\eta_r)\}_{r=0}^\infty$ is bounded. By construction, there exists $s(\eta) \in \partial_\varepsilon f(x(\eta))$ such that $\|s(\eta)\| \leq \eta$. The boundedness of $\{x(\eta_r)\}$ implies that this sequence has an accumulation point, say $x(0)$. If follows from the closedness of the map $x \to \partial_\varepsilon f(x)$ that $0 \in \partial_\varepsilon f(x(0))$. Thus $\lim_{r \to \infty} f(x(\eta_r)) = f(x(0)) \leq \inf f(x) + \varepsilon$, as required. $\square$

In recent years many algorithms similar in spirit to the $L$-algorithm were suggested, which differ from it in approximate line search rules, in tests used at Step 5, and in stopping criteria [52, 101, 55]. R. Mifflin [55] extended the $L$-algorithms to a much broader class of nonsmooth functions and suggested modifications for solving constrained problems. It is also worth noting that $\varepsilon$-subgradient methods were treated in recent papers by V. F. Demyanov, L. V. Vasilyev and E. A. Nurminski.

## 2.6 An Extension of the Subgradient Method to a Class of Nonconvex Functions. Stochastic Versions and Stability of the Method

In order to extend nonmonotone variants of the subgradient method to nonconvex functions, one has to find a condition which would replace, in a certain sense, the obvious property of convex functions that the angle between the gradient and the direction towards the minimum is obtuse. Such a condition is used in the theorem proved by L. G. Bazhenov [5].

**Theorem 2.18.** *Let $f$ be an almost differentiable function and let $x^*$ be its local minimum point such that*

$$f(x^*) = \min_{x \in S_r} f(x),$$

*where*

$$S_r = \{x : \|x - x^*\| \le r\}.$$

*Suppose that for any $\varepsilon$ satisfying $0 < \varepsilon < r$ one has*

$$\inf_{x \in S_r \setminus S_\varepsilon} (g_f(x), x - x^*) > 0.$$

*If $\|x_0 - x^*\| \le r$ then the sequence $\{x_k\}_{k=0}^{\infty}$ generated by*

$$x_{k+1} = \begin{cases} \bar{x}_{k+1} & \text{if } \bar{x}_{k+1} \in S_r, \\ x_0 & \text{if } \bar{x}_{k+1} \notin S_r, \end{cases}$$

*where*

$$\bar{x}_{k+1} = x_k - h_k \frac{g_f(x_k)}{\|g_f(x_k)\|},$$

$$h_k > 0, \ \lim_{k \to \infty} h_k = 0, \ \sum_{k=0}^{\infty} h_k = +\infty,$$

*converges to $x^*$.*

In [27] the notion of *stochastic subgradient* was introduced and used for elaborating a random search method which is a stochastic analogue of the subgradient method. Further developed in the works of Yu. M. Ermoliev, E. A. Nurminski and others (see [25]), the stochastic subgradient method became an efficient tool for solving manifold stochastic programming problems. To complete our review we shall briefly discuss its basic ideas. The method belongs to the broad class of random search methods, i.e. iterative algorithms that proceed along directions resulting from random (pseudorandom) events, contrary to deterministic procedures in ordinary gradient methods (if various numerical errors are neglected).

The *stochastic subgradient method* for minimizing a convex function $f$ is defined by the formula:

$$x_{k+1} = x_k - h_k(x_k)\, g_\omega(x_k), \quad k = 0, 1, \ldots,$$

where $h_k$ is a stepsize coefficient at the $k$-th iteration, and $g_\omega(x_k)$ is a random vector having the mathematical expectation equal to a subgradient of $f$ at $x_k$. For simplicity we assume that the probabilistic characteristics of the vector $g_\omega(x_k)$ are fully determined by the point $x_k$ and do not depend on the past of the search process (this requirement is not essential and one can prove convergence in more general cases).

Let $x^*$ be the unique minimum point for $f$. We have the following theorem [27].

**Theorem 2.19.** *Let the following conditions be satisfied:*

(i) $\displaystyle\sum_{k=0}^{\infty} h_k(x_k) = +\infty; \quad h_k(x_k) > 0;$

(ii) $\displaystyle\sum_{k=0}^{\infty} h_k^2(x_k) < \infty;$

(iii) $E\{\|g_\omega(x_k)\|^2\} \leqq c, \quad k = 0, 1, \ldots,$     (*E denotes the expectation*).

*Then with probability one*

$$\lim_{k\to\infty} \|x_k - x^*\| = 0.$$

*Proof.* The proof of convergence is based on properties of random sequences called supermartingales. A sequence of random variables $\{y_k\}_{k=1}^{\infty}$ is a *supermartingale*, if

$$E\{y_n/y_{n-1}, \ldots, y_0\} \leqq y_{n-1},$$

where $E\{y_n/y_{n-1}, \ldots, y_1\}$ denotes the conditional expectation of $y_n$ with respect to $y_{n-1}, \ldots, y_0$. This concept generalizes the notion of monotonic sequences.

The following convergence theorem holds for supermartingales [19]: if $E\{|y_n|\} \leqq c < +\infty$ then with probability one the limit

$$\lim_{n\to\infty} y_n = y^*,$$

exists, and $E\{|y^*|\} < +\infty$.

Let us study the sequence $\{E\{\|x_k - x^*\|^2\}\}_{k=0}^{\infty}$. Observe that

$$\|x_{k+1} - x^*\|^2 = \|x_k - h_k(x_k)\, g_\omega(x_k) - x^*\|^2$$
$$= \|x_k - x^*\|^2 - 2h_k(x_k)(g_\omega(x_k), x_k - x^*) + h_k^2(x_k)\|g_\omega(x_k)\|^2,$$
$$E\{\|x_{k+1} - x^*\|^2/x_k\} = \|x_k - x^*\|^2 - 2h_k(x_k)(E\{g_\omega(x_k)/x_k\}, x_k - x^*)$$
$$+ h_k^2 E\{\|g_\omega(x_k)\|^2/x_k\}.$$

Since $E\{g_\omega(x_k)/x_k\} = g_f(x_k)$ and $(g_f(x_k), x_k - x^*) \geqq 0$, we have

$$E\{\|x_k - x^*\|^2/x_k\} \leqq \|x_k - x^*\|^2 + c\, h_k^2(x_k). \tag{2.42}$$

Consider the random variable $z_k = \|x_k - x^*\|^2 + c\sum_{s=k}^{\infty} h_s^2(x_s)$. Inequality (2.42) is equivalent to the inequality $E\{z_{k+1}/z_k, \ldots, z_1\} \leqq z_k$. Consequently, $\{z_k\}_{k=1}^{\infty}$ is a supermartingale and converges almost surely to a certain limit $z^*$. Since $\lim_{k\to\infty} \sum_{s=k}^{\infty} h_s^2(x_s) = 0$, the sequence $\{\|x_k - x^*\|^2\}$ also converges to $z^*$ with probability one. We shall prove by contradiction that the limit $z^*$ is equal to zero. If this were not true, then one could find $\varepsilon > 0$ and $\delta > 0$ such that with probability $\delta$ one would have $\| x_k - x^* \| \geqq \varepsilon$ for all sufficiently large $k$, whence

$$\sum_{k=0}^{\infty} h_k(x_k)(E\{g_\omega(x_k)/x_k\}, x_k - x^*) = + \infty$$

with probability $\delta$, which in turn would yield

$$E\left\{\sum_{k=0}^{\infty} h_k(x_k)(g_\omega(x_k), x_k - x^*)\right\} = + \infty.$$

However, this would contradict the relation

$$E\{\| x_{k+1} - x^* \|^2\} = | x_0 - x^* |^2 - 2E\left\{\sum_{s=0}^{k} h_s(x_s)(g_\omega(x_s), x_s - x^*)\right\}$$

$$+ E\left\{\sum_{s=0}^{k} h_s^2(x_s)\| g_\omega(x_s) \|^2\right\}.$$

The proof is complete.  $\square$

The stochastic subgradient 'method has many practical applications, especially to the solution of multistage problems of stochastic programming [27, 78].

Condition (ii) of the theorem is stronger than the condition $h_k \to 0$ used in the substantiation of the subgradient method, but if may be fulfilled without special difficulties (e.g., $h_k = 1/(k + 1)$ satisfies our requirements).

A similar condition was used by M. A. Shepilov [73] for proving stability of the subgradient method with respect to small errors in the computation of $x_k$ and $g_f(x_k)$. He obtained the following result.

**Theorem 2.20.** *If the function f is convex on $E_n$ and the set $M^*$ of its minima is nonempty, then for any $x_0$ a sequence $\{x_k\}_{k=0}^{\infty}$ generated by*

$$x_{k+1} = x_k - h_k \frac{g_f(\bar{x}_k)}{\| g_f(\bar{x}_k) \|}, \quad k = 0, 1, 2, \ldots,$$

*where*

$$\| \bar{x}_k - x_k \| \leqq \delta_k, \lim_{k \to \infty} \delta_k = 0,$$

$$h_k > 0, \quad \sum_{k=0}^{\infty} h_k \delta_k < \infty, \quad \sum_{k=0}^{\infty} h_k^2 < \infty, \quad \sum_{k=0}^{\infty} h_k = \infty.$$

*converges to a point $x^* \in M^*$.*

The above theorem together with the method for approximate computation of subgradients, which was described in Sect. 1.3, give grounds for constructing a "general-purpose" algorithm for minimizing convex functions which does not require calculation of subgradients (provided that the functions can be evaluated at any point with any required accuracy).

# 3. Gradient-type Methods with Space Dilation

## 3.1 Heuristics of Methods with Space Dilation

The analysis of the subgradient method has shown that improvements only in the stepsize rules cannot, in general, significantly accelerate convergence if at each iteration the algorithm proceeds in the direction negative to that of the gradient. Indeed, slow convergence is due to the fact that the gradient is almost perpendicular to the direction towards the minimum. In such circumstances the reduction of the distance to the minimum is much smaller than the stepsize, and therefore the stepsizes cannot diminish too rapidly, if we want to guarantee convergence to a minimum.

On the other hand, there is a simple way of changing the angles between the gradient and the direction towards the minimum, namely the application of linear nonorthogonal space transformations. Thus an idea arises to construct at each iteration a certain linear operator changing the metric of the space, and to use the direction negative to that of the gradient in the transformed space. Such a direction may significantly differ from the direction negative to that of the gradient in the original space.

How should the operators of space transformation be constructed? Recalling well-known quasi-Newton algorithms for minimizing twice continuously differentiable functions, we see that the basic idea of the variable metric methods consists in obtaining, in one way or another, a matrix close to the inverse of the Hessian (or its positive multiple) at the minimum, i.e. one uses a quadratic approximation to the objective function and imitates the Newton-Raphson method without calculating the second derivatives [69]. This approach does not apply to nonsmooth functions for fundamental reasons. For instance, the Hessian of a piecewise linear function vanishes almost everywhere, and its inverse simply does not exist. Therefore, other ideas should be employed for constructing gradient-type methods with space transformation for a sufficiently broad class of nonsmooth functions.

As indicated above, the basic factor that slows down the subgradient method with stepsizes determined off line, is that the cosine of the angle between the gradient and the direction towards the minimum is close to zero. Were it possible to substantially reduce the component of the gradient that is orthogonal to the direction towards the minimum, while leaving the component that is parallel to that direction almost unchanged, the resulting vector would define a better search direction than the gradient, and so one could expect on acceleration of convergence, provided an appropriate stepsize rule is used. But how can we find a direction providing a significant decrease of the function, or directions orthogonal to it? Let us observe that in most difficult cases it is the direction of the gradient that is almost orthogonal to the direction towards the minimum. Assuming such an orthogonality, we should try to reduce at subsequent iterations the components of the gradient that are parallel to the latest gradient. This can be done by performing space dilation in the direction of the gradient. So, from rather informal heuristic considerations we have derived the idea of the subgradient method with space dilation along the gradient. A detailed description and substantiation of the method will be given below.

Another class of gradient-type methods with space dilation (performed in this case along the direction of the difference of two successive gradients) is connected with successive space transformations, which aim, in general, at transforming acute cones of descent directions into wider ones. In this way one obtains an almost monotone method, which proved efficient for a broad class of practical problems. This method is discussed in Sects. 3.6 and 3.7.

In Sect. 3.8 a group of methods for minimizing convex functions is considered, in which supporting hyperplanes are used as cutting planes and the regions resulting from successive out-offs are approximated with ellipsoids. Space dilation is used for transforming ellipsoids into spheres (for symmetrization). An important special case of such algorithms, which belong to the family of algorithms with space dilation along the gradient was first suggested in [103], and independently in [87]. All these algorithms are of special theoretical and practical interest, since they ensure convergence of function values with the speed of a geometrical progression of a ratio depending only on the dimension of the space.

## 3.2 Operators of Space Dilation

Let a vector $\xi \in E_n$, $\| \xi \| = 1$, and a number $\alpha \geq 0$ be fixed. Every vector $x \in E_n$ may be represented as follows:

$$x = \gamma_\xi(x)\, \xi + d_\xi(x), \tag{3.1}$$

where

$$(x, d_\xi(x)) = 0. \tag{3.2}$$

From (3.1) and (3.2) we obtain $\gamma_\xi(x) = (x, \xi)$ and $d_\xi(x) = x - (x, \xi)\, \xi$.

**Definition.** *An operator $R_\alpha(\xi)$ which transforms a vector $x$ of the form* (3.1) *into*

$$R_\alpha(\xi)\, x = \alpha\, \gamma_\xi(x)\, \xi + d_\xi(x),$$

*is called an operator of space dilation along direction $\xi$ with coefficient $\alpha$.*

It follows from the above definition that the following statements are true:

1) $R_\alpha(\xi)\, x = \alpha(x,\,\xi)\,\xi + [x - (x,\,\xi)\,\xi] = (\alpha - 1)(x,\,\xi)\,\xi + x.$      (3.3)
2) The operator $R_\alpha(\xi)$ is linear and symmetric,

$$(R_\alpha(\xi)\, x,\, y) = (\alpha - 1)(x,\,\xi)(y,\,\xi) + (x,\, y) = (x,\, R_\alpha(\xi)\, y).$$

3) $R_{\alpha\beta}(\xi) = R_\alpha(\xi)\, R_\beta(\xi).$
4) For $\alpha > 0$ one has $R_\alpha(\xi)\, R_{1/\alpha}(\xi) = R_1(\xi) = I$, where $I$ is the identity matrix.
5) The operator $R_0(\xi)$ is the operator of projection on the subspace orthogonal to $\xi$,

$$R_0(\xi)\, x = d_\xi(x).$$

6) The operator $R_\alpha(\xi)$ has for $n \geq 2$ two eigenvalues $\lambda_1 = \alpha$, and $\lambda_2 = 1$, the first one corresponding to the subspace of eigenvectors generated by $\xi$, and the second to the subspace of eigenvectors orthogonal to $\xi$.
7) If the coordinates of the vector $\xi$ in a certain orthonormal basis $s = \{e_1, e_2, \ldots, e_n\}$ are equal to $\xi_1, \xi_2, \ldots, \xi_n$, then it follows from (3.3) that the operator $R_\alpha(\xi)$ is represented in this coordinate system by the matrix $R_\alpha(\xi)$ with entries $\{r_{ij}\}$ defined as follows:

$$r_{ij} = (R_\alpha(\xi)\, e_i,\, e_j) = \begin{cases} (\alpha - 1)\,\xi_i\,\xi_j & \text{for } i \neq j, \\ (\alpha - 1)\,\xi_i^2 + 1 & \text{for } i = j. \end{cases}$$

8) By (3.3), the computation of a vector $R_\alpha(\xi)\, x$ requires $(2\,n + 1)$ multiplications, and the computation of matrices of the form $R_\alpha(\xi)\, A$ or $AR_\alpha(\xi)$ for a given matrix $A$ requires $n\,(2\,n + 1)$ multiplications.
9) For any vector $x \in E_n$ one has

$$\| R_\alpha(\xi)\, x \| = \sqrt{\| x \|^2 + (\alpha^2 - 1)(x,\,\xi)^2}\,, \tag{3.4}$$

since from (3.3) we obtain

$$\begin{aligned}
\| R_\alpha(\xi)\, x \|^2 &= (R_\alpha(\xi)\, x,\, R_\alpha(\xi)\, x) \\
&= (x + (\alpha - 1)(x,\,\xi)\,\xi,\, x + (a - 1)(x,\,\xi)\,\xi) \\
&= \| x \|^2 + 2(\alpha - 1)(x,\,\xi)^2 + (\alpha - 1)^2(x,\,\xi)^2 \\
&= \| x \|^2 + (\alpha^2 - 1)(x,\,\xi)^2.
\end{aligned}$$

10) $R_\alpha(\xi)$ has the matrix representation

$$R_\alpha(\xi) = I + (\alpha - 1)\,\xi\,\xi^T.$$

Indeed,

$$(I + (\alpha - 1)\, \xi\, \xi^T)\, x = x + (\alpha - 1)\, \xi\, (\xi, x), \tag{3.5}$$

which is exactly the defining relation (3.3).

## 3.3 The Subgradient Method with Space Dilation in the Direction of the Gradient

We shall consider a class of algorithms for minimizing convex functions, in which movement in the direction of the subgradient is combined with space dilation along this direction. Algorithms from this class will be called *subgradient methods with space dilation along the gradient (SDG algorithms,* SDG for "Space Dilation along the Gradient").

Assume that we have a finite process for computing a subgradient $g_f(x)$ of a convex objective function $f$ at any point $x \in E_n$, rules for determining sequences of positive reals $\{h_k\}$ and $\{\alpha_k\}$, $k = 1, 2, \ldots$, (stepsizes and space dilation coefficients), a starting point $x_0$, and an initial nonsingular matrix $B_0 = A_0^{-1}$ (e.g. $B_0 = I$). Under these assumptions we shall describe an infinite iterative algorithm, which has the $(k + 1)$-st iteration, $k = 0, 1, 2, \ldots$, defined as follows:

1) Evaluate $g_f(x_k)$ (if $g_f(x_k) = 0$ the computations are stopped, since then $x_k$ is the optimal point).
2) Set

$$g_{\varphi_k}(y_k) = B_k^* g_f(x_k) = \tilde{g}_k, \tag{3.6}$$

where $\varphi_k(y) = f(B_k y)$, $y_k = A_k x_k$, $A_k = B_k^{-1}$ and $B_k^*$ is the adjoint operator to $B_k$. By formula (3.6) we calculate a subgradient of the function $\varphi_k(y) = f(A_k^{-1} y)$, which is obtained from $f$ by the linear transformation of variables $y = A_k x$. This formula is a special case of a general formula for the supporting functional of a convex functional defined on a Banach space which is a linear transformation of another Banach space [68]. The formula may be also easily derived from the definition of the subgradient. At those points at which $g_f(x)$ is nonunique, formula (3.6) should be read as a single-valued mapping from the set $G_f(x)$ into the set $G_{\varphi_k}(y)$ (if the operator $B_k$ is nonsingular then this mapping is surjective and invertible).

Next compute

$$3)\quad \xi_{k+1} = \frac{g_{\varphi_k}(y_k)}{\| g_{\varphi_k}(y_k) \|} = \frac{\tilde{g}_k}{\| \tilde{g}_k \|}, \tag{3.7}$$

$$4)\quad h_{k+1},$$

$$5)\quad \alpha_{k+1},$$

$$6)\quad x_{k+1} = x_k - h_{k+1}\, B_k\, \xi_{k+1}. \tag{3.8}$$

Relation (3.8) is obtained from the formula $A_k\,x_{k+1} = y_k - h_{k+1}\,\xi_{k+1}$, which describes one iteration of the subgradient method for minimizing $\varphi_k$ in the transformed space, by premultiplying it with the operator $B_k$ to map both sides into the original space.

7) Calculate

$$B_{k+1} = A_{k+1}^{-1} = B_k\,R_{1/\alpha_{k+1}}(\xi_{k+1}). \tag{3.9}$$

Formula (3.9) defines the operator $B_{k+1}$, which is the inverse of the space transformation operator

$$A_{k+1} = R_{\alpha_{k+1}}(\xi_{k+1}) \ldots R_{\alpha_1}(\xi_1)\,A_0$$

resulting from successive space dilations in the directions of the normed subgradients $\xi_1, \xi_2, \ldots, \xi_{k+1}$ with coefficients $\alpha_1, \ldots, \alpha_{k+1}$. Thus

$$B_{k+1} = A_{k+1}^{-1} = A_k^{-1}\,R_{\alpha_{k+1}}^{-1}(\xi_{k+1}) = B_k\,R_{1/\alpha_{k+1}}(\xi_{k+1}).$$

8) Proceed to the $(k+2)$-nd iteration.

At each iteration the SDG method involves more operations than the subgradient method. The most laborious ones are the operations 2), 6) and 7), each of them requiring a number of arithmetic operations of order $n^2$, including about $4\,n^2$ multiplications. Moreover, additional memory is required for storing matrices $B_k$.

Let us observe that if $B_0 = I$ and $\alpha_k = 1$ for $k = 1, 2, \ldots$, then the SDG algorithm becomes equivalent to the subgradient method. But the SDG method is more flexible since its operation depends on two sequences $\{\alpha_k\}$ and $\{h_k\}$. Using various rules for determining those sequences, we can modify the algorithm in many ways and choose the most efficient version for a given class of problems.

The methods with space dilation had been initially developed for minimizing convex functions. However, it soon turned out that some of their versions may be used for minimizing (at least locally) almost differentiable functions (with subgradients replaced by almost-gradients).

In our further considerations, when studying convergence of SDG algorithms, we shall always make clear whether we deal with the class of almost differentiable functions or the subclass of convex functions. The reader should interpret the term "generalized gradient" as the almost-gradient or the subgradient, respectively.


## 3.4 Convergence of Algorithms with Space Dilation


We shall show that certain SDG algorithms improve the function values with the speed of a geometrical progression with a ratio depending on properties of the objective function that are invariant with respect to nonsingular linear space transformations. Moreover, for a special SDG algorithm for convex minimization this ratio depends only on the dimension of the domain of the function. This algorithm is described in Sect. 3.8.

The proofs of the results to follow are based on the analysis of the behavior of the norm of the gradient of the objective function in the transformed space.

**Theorem 3.1.** *Let the SDG method* (3.6)−(3.9) *be applied to the minimization of an almost differentiable function f and let it generate a sequence* $\{x_k\}_{k=0}^{\infty}$. *Suppose that there exist positive numbers d, $\alpha^*$ and $\delta$ such that for $k = 0, 1, \ldots$ the following conditions are satisfied:*

1) $\|g_f(x_k)\| \le d$,

2) $1 + \delta \le \alpha_k \le \alpha^*$.

*Then one can find a subsequence* $\{x_{k_p}\}_{p=0}^{\infty}$, $k_p < k_{p+1}$, *and a constant $c > 0$ such that*

$$\|\tilde{g}_{k_p}\| < c \left( \prod_{j=1}^{k_p} \alpha_j \right)^{-1/n} \qquad \text{for } p = 0, 1, \ldots .$$

*Proof.* Let us express the matrix $A_k$ as a product of an orthogonal matrix $O_k$ and a symmetric positive definite matrix $S_k$ (the polar factorization):

$$A_k = O_k S_k; \quad A_0 = I.$$

Since $A_k = R_{\alpha_k}(\xi_k) \ldots R_{\alpha_1}(\xi_1)$ and the product of the eigenvalues of a matrix is equal to its determinant, the product of the eigenvalues of the matrix $S_k$ is equal to

$$\det S_k = \det A_k = \prod_{j=1}^{k} \alpha_j .$$

Let $\{\lambda_i^{(k)}\}_{i=1}^{n}$ be the eigenvalues of $S_k$, $\lambda_1^{(k)} \le \lambda_2^{(k)} \le \ldots \le \lambda_n^{(k)}$, and let $\{e_i^{(k)}\}_{i=1}^{n}$ be the corresponding orthonormal system of eigenvectors. We have

$$S_k e_i^{(k)} = \lambda_i^{(k)} e_i^{(k)}; \quad \prod_{i=1}^{n} \lambda_i^{(k)} = \prod_{j=1}^{k} \alpha_j; \quad \lambda_n^{(k)} \ge \left( \prod_{j=1}^{k} \alpha_j \right)^{1/n} . \qquad (3.10)$$

Moreover, $\lambda_i^{(k)} \ge 1$, since $\alpha_j > 1, j = 1, \ldots, k$, and thus

$$\| S_k e_i^{(k)} \| = \| A_k e_i^{(k)} \| \ge 1 .$$

Let $O_k e_i^{(k)} = \bar{e}_i^{(k)}$. The system of vectors $\{\bar{e}_i^{(k)}\}_{i=1}^{n}$ is also orthonormal, because $O_k$ is orthogonal.

Let us write $g_f(x_k)$ as follows:

$$g_f(x_k) = \sum_{i=1}^{n} g_i^{(k)} e_i^{(k)} .$$

Since $\| g_f(x_k) \| \le d$,

$$|g_i^{(k)}| \le d; \quad i = 1, 2, \ldots, n; \quad k = 1, 2, \ldots . \qquad (3.11)$$

We shall prove the theorem by contradiction.

Suppose that the assertion is false. Then for arbitrarily large $c$ one can find $k(c)$ such that for all $k > k(c)$

$$\|\tilde{g}_k\| = \|B_k^* g_f(x_k)\| = \left\| \sum_{i=1}^{n} \frac{g_i^{(k)}}{\lambda_i^{(k)}} \bar{e}_i^{(k)} \right\| \geq c \left( \prod_{j=1}^{k} \alpha_j \right)^{-1/n}. \tag{3.12}$$

Thus, for any $k > k(c)$ there exists $i_k^*$, $1 \leq i_k^* \leq n$, such that

$$\frac{|g_{i_k^*}^{(k)}|}{\lambda_{i_k^*}^{(k)}} \geq \frac{c \left( \prod\limits_{j=1}^{k} \alpha_j \right)^{-1/n}}{\sqrt{n}},$$

whence

$$\lambda_{i_k^*}^{(k)} \leq \frac{\sqrt{n}\, d}{c} \left( \prod_{j=1}^{k} \alpha_j \right)^{1/n}.$$

Next,

$$\lambda_n^{(k)} \max_{1 \leq i \leq n} \lambda_i^{(k)} \geq \left( \frac{\prod\limits_{j=1}^{k} \alpha_j}{\lambda_{i_k^*}^{(k)}} \right)^{1/(n-1)} \geq \left( \frac{c}{\sqrt{n}\, d} \right)^{1/(n-1)} \left( \prod_{j=1}^{k} \alpha_j \right)^{1/n}$$

$$= \tilde{c} \left( \prod_{j=1}^{k} \alpha_j \right)^{1/n}, \quad \tilde{c} = \left( \frac{c}{\sqrt{n}\, d} \right)^{1/(n-1)}.$$

Let

$$e_n^{(k+1)} = \sum_{i=1}^{n} d_i^{(k)} e_i^{(k)}; \quad \sum_{i=1}^{n} (d_i^{(k)})^2 = 1.$$

Then

$$A_{k+1} e_n^{(k+1)} = \lambda_n^{(k+1)} \bar{e}^{(k+1)} = R_{\alpha_{k+1}}(\xi_{k+1}) \left( \sum_{i=1}^{n} \lambda_i^{(k)} d_i^{(k)} \bar{e}_i^{(k)} \right).$$

Set $r_k = \min \{r: \lambda_n^{(k)}/\lambda_r^{(k)} \leq \alpha_{k+1} c, 1 \leq r \leq n\}$ and express $y = \sum_{i=1}^{n} \lambda_i^{(k)} d_i^{(k)} \bar{e}_i^{(k)}$ as $y = y_1 + y_2$, where

$$y_1 = \sum_{i=1}^{r_k-1} \lambda_i^{(k)} d_i^{(k)} \bar{e}_i^{(k)},$$

$$y_2 = \sum_{i=r_k}^{n} \lambda_i^{(k)} d_i^{(k)} \bar{e}_i^{(k)}.$$

Let us estimate $\| R_{\alpha_{k+1}}(\xi_{k+1}) y \|$. To this end observe that

$$\xi_{k+1} = \frac{\tilde{g}_k}{\|\tilde{g}_k\|} = \sum_{i=1}^{n} \mu_i^{(k)} \bar{e}_i^{(k)},$$

where $\mu_i^{(k)} = g_i^{(k)}/\lambda_i^{(k)} \|\tilde{g}_k\|$. From (3.11) and (3.12) we obtain

$$|\mu_i^{(k)}| = \frac{d\left(\prod_{j=1}^{k} \alpha_j\right)^{1/n}}{\lambda_i^{(k)} c}.$$

For $i \geq r_k$ we have

$$|\mu_i^{(k)}| \leq \frac{d\left(\prod_{j=1}^{k} \alpha_j\right)^{1/n} \alpha_{k+1}}{\lambda_n^{(k)}} \leq \frac{d\alpha^*}{\tilde{c}}.$$

Thus

$$\| R_{\alpha_{k+1}}(\xi_{k+1})\, y_1 \| \leq \alpha_{k+1} \| y_1 \| \leq \lambda_n^{(k)} \frac{1}{c} \left\| \sum_{i=1}^{r_k-1} d_i^{(k)}\, \bar{e}_i^{(k)} \right\| \leq \frac{\lambda_n^{(k)}}{c}.$$

$$\leq \lambda_n^{(k)} \left[ 1 + (\alpha^* - 1)\, \frac{d\alpha^*}{\tilde{c}} \right],$$

$$\| R_{\alpha_{k+1}}(\xi_{k+1})\, y_2 \| \leq \| y_2 \| + (\alpha_{k+1} - 1)\, |(y_2, \xi_{k+1})|$$

$$\lambda_n^{(k+1)} = \| R_{\alpha_{k+1}}(\xi_{k+1})\, y \| \leq \lambda_n^{(k)} \left[ 1 + \frac{1}{c} + (\alpha^* - 1)\, \frac{d\alpha^*}{\tilde{c}} \right]. \qquad (3.13)$$

By (3.13), for any $\varepsilon_0 > 0$ one can always find $c$ so large that $\lambda_n^{(k+1)} \leq (1 + \varepsilon_0)\, \lambda_n^{(k)}$ for all $k > k(c)$. But if we choose $\varepsilon_0 < (1 + \delta)^{1/n} - 1$ then for sufficiently large $k$ we obtain $\lambda_n^{(k)} < (1 + \varepsilon_0)^k < (1 + \delta)^{k/n} \leq (\prod_{j=1}^{k} \alpha_j)^{1/n}$, which contradicts (3.10). This proves the theorem. $\square$

A similar result may be obtained for the rate of convergence of the sequence $v_k = \min_{1 \leq r \leq k} \|\tilde{g}_r\|$. This result will be used later to estimate the rate of convergence of the function value "records" in SDG algorithms (the "record" at the $k$-th iteration equals $\min_{1 \leq r \leq k} f(x_r)$).

**Theorem 3.2.** *Under the conditions of Theorem 3.1 assume that* $\alpha_k = \alpha > 1$ *for all $k$. Then*

$$v_k \leq \frac{d\sqrt{k(\alpha^2 - 1)}}{\sqrt{\alpha^{2k/n} - 1}} \quad \textit{for } k = 1, 2, \dots$$

*Proof.* We shall use the notation introduced in the proof of Theorem 3.1. We have

$$\xi_{k+1} = \frac{\tilde{g}_k}{\|\tilde{g}_k\|} = \sum_{i=1}^{n} \mu_i^{(k)}\, \bar{e}_i^{(k)}, \quad k = 1, 2, \dots,$$

$$\mu_i^{(k)} = \frac{g_i^{(k)}}{\lambda_i^{(k)} \|\tilde{g}_k\|}, \quad i = 1, 2, \dots, n, \quad \sum_{i=1}^{n} (g_i^{(k)})^2 \leq d^2. \qquad (3.14)$$

Suppose that

$$v_k = \min_{1 \le r \le k} \|\tilde{g}_k\| > \frac{d\sqrt{k(\alpha^2 - 1)}}{\sqrt{\alpha^{2k/n} - 1}}$$

for some $k$. Then for $1 \le r \le k$ and $1 \le i \le n$ (3.14) yields

$$|\mu_i^{(r)}| \le \frac{|g_i^{(r)}|\sqrt{\alpha^{2k/n} - 1}}{\lambda_i^{(r)} d\sqrt{k(\alpha^2 - 1)}}. \tag{3.15}$$

Using this relation we shall estimate $(\lambda_n^{(r+1)})^2 - (\lambda_n^{(r)})^2$ for $1 \le r \le k - 1$ ($\lambda_n^{(r)}$ is the largest eigenvalue of the matrix $S_r$). For any vector $a$ such that $\|a\| = 1$ we have

$$a = \sum_{i=1}^{n} a_i e_i^{(r)}, \quad \sum_{i=1}^{n} a_i^2 = 1,$$

$$A_r a = \sum_{i=1}^{n} a_i \lambda_i^{(r)} \bar{e}_i^{(r)}, \tag{3.16}$$

$$\lambda_n^{(r+1)} = \max_{\|a\|=1} \|A_{r+1} a\| = \max_{\|a\|=1} \|R_\alpha(\xi_{r+1}) A_r a\|. \tag{3.17}$$

Recall formula (3.4):

$$\|R_\alpha(\xi) x\| = \sqrt{\|x\|^2 + (\alpha^2 - 1)(x, \xi)^2}.$$

Setting $x = A_r a$, we obtain from (3.15)–(3.17) the inequality

$$[\lambda_n^{(r+1)}]^2 = \max_{\|a\|=1} \left[ \sum_{i=1}^{n} a_i^2 (\lambda_i^{(r)})^2 + (\alpha^2 - 1)\left( \sum_{i=1}^{n} a_i \lambda_i^{(r)} \mu_i^{(r)} \right)^2 \right]$$

$$\le \max_{\|a\|=1} \left[ (\lambda_n^{(r)})^2 + (\alpha^2 - 1)\left( \sum_{i=1}^{n} |a_i g_i^{(r)}| \right)^2 \frac{\alpha^{2k/n} - 1}{k d^2 (\alpha^2 - 1)} \right]$$

$$< [\lambda_n^{(r)}]^2 + \frac{\alpha^{2k/n} - 1}{k}, \quad r = 0, 1, \ldots, k.$$

Since $\lambda_n^{(0)} = 1$, one has $[\lambda_n^{(k)}]^2 < \alpha^{2k/n}$, which contradicts the inequality $\lambda_n^{(k)} \ge \alpha^{k/n}$. The proof is complete. $\square$

To estimate the rate of convergence of function values in the SDG method we shall need estimates of the distance in the transformed space from the current point to the optimal set. Such estimates can be easily derived for some versions of the SDG method.

**Theorem 3.3.** *Let $f$ be an almost differentiable function defined on a certain sphere $S_d = \{x: \|x - x^*\| \le d\}$ around the local minimum point $x^*$ of $f$, and let for all $x \in S_d$ the almost-gradient $g(x)$ satisfy the inequalities*

$$N[f(x) - f(x^*)] \le (g(x), x - x^*) \le M[f(x) - f(x^*)], \tag{3.18}$$

*where $M > N > 0$. If in the SDG method we set*

1) $\quad x_0 \in S_d$;

2) $\quad h_{k+1} = \dfrac{2MN}{M+N} \dfrac{f(x_k) - f(x^*)}{\|\tilde{g}_k\|}$; $\qquad\qquad\qquad\qquad$ (3.19)

3) $\quad 1 < \alpha_{k+1} = \alpha \leq \dfrac{M+N}{M-N}, \quad k = 0, 1, 2, \ldots,$ $\qquad\qquad$ (3.20)

*then*

$$\|A_k(x_k - x^*)\| \leq d \quad \text{for} \quad k = 0, 1, 2, \ldots .$$

*Proof.* We shall prove the theorem by induction. Let

$$\|A_p(x_p - x^*)\| \leq d$$

for some $p$. We shall estimate

$$' \|A_{p+1}(x_{p+1} - x^*)\|.$$

Denote $A_k(x_k - x^*) = z_k$. We have

$$\|z_{p+1}\|^2$$

$$= \|R_{\alpha_{p+1}}(\xi_{p+1})(z_p - h_{p+1}\,\xi_{p+1})\|^2$$

$$= \|R_{\alpha_{p+1}}(\xi_{p+1})\,z_p\|^2 - 2h_{p+1}(R_{\alpha_{p+1}}(\xi_{p+1})\,z_p, R_{\alpha_{p+1}}(\xi_{p+1})\,\xi_{p+1}) + \alpha_{p+1}^2\,h_{p+1}^2$$

$$= \left[1 + (\alpha_{p+1}^2 - 1)\frac{(z_p, \xi_{p+1})^2}{\|z_p\|^2}\right]\|z_p\|^2 - 2h_{p+1}\,\alpha_{p+1}^2\,(z_p, \xi_{p+1}) + \alpha_{p+1}^2\,h_{p+1}^2$$

$$= \|z_p\|^2 + \alpha_{p+1}^2\,[(z_p, \xi_{p+1}) - h_{p+1}]^2 - (z_p, \xi_p)^2. \qquad\qquad (3.21)$$

Observe that

$$(z_p, \xi_{p+1}) = (A_p(x_p - x^*), \xi_{p+1}) = (x_p - x^*, A_p^*\,\xi_{p+1})$$

$$= \frac{1}{\|\tilde{g}_p\|}\,(g_f(x_p), x_p - x^*).$$

Let us consider two cases.

*Case (a):* $(z_p, \xi_{p+1}) \leq h_{p+1}$.

Using the inequality

$$h_{p+1} = \frac{[f(x_p) - f(x^*)]\,2MN}{\|\tilde{g}_p\|\,(M+N)} \leq \frac{(g_f(x_p), x_p - x^*)\,2M}{\|\tilde{g}_p\|\,(M+N)}$$

$$= (z_p, \xi_{p+1})\,\frac{2M}{M+N},$$

which results from (3.18) and (3.19), we see that

$$\alpha_{p+1}^2 \left[(z_p, \xi_{p+1}) - h_{p+1}\right]^2 - (z_p, \xi_{p+1})^2$$

$$\leq \alpha_{p+1}^2 \left[(z_p, \xi_{p+1}) - (z_p, \xi_{p+1})\frac{2M}{M+N}\right]^2 - (z_p, \xi_{p+1})^2$$

$$\leq \left[\left(\frac{M+N}{M-N}\right)^2 \left(1 - \frac{2M}{M+N}\right)^2 - 1\right] (z_p, \xi_{p+1})^2 = 0.$$

*Case (b):* $(z_p, \xi_{p+1}) \geq h_{p+1}$.

In a similar way we obtain

$$h_{p+1} = \frac{[f(x_p) - f(x^*)]\, 2MN}{\|\tilde{g}_p\|\, (M+N)} \geq \frac{(g_f(x_p), x_p - x^*)\, 2N}{\|\tilde{g}_p\|\, (M+N)}$$

$$= (z_p, \xi_{p+1})\frac{2N}{M+N},$$

and

$$\alpha_{p+1}^2 \left[(z_p, \xi_{p+1}) - h_{p+1}\right]^2 - (z_p, \xi_{p+1})^2$$

$$\leq \alpha_{p+1}^2 \left[(z_p, \xi_{p+1}) - (z_p, \xi_{p+1})\frac{2N}{M+N}\right]^2 - (z_p, \xi_{p+1})^2$$

$$\leq \left[\left(\frac{M+N}{M-N}\right)^2 \left(1 - \frac{2N}{M+N}\right)^2 - 1\right] (z_p, \xi_{p+1})^2 = 0.$$

Hence, in view of inequality (3.2) $\|z_{p+1}\| \leq \|z_p\|$. Consequently, $\|A_k(x_k - x^*)\| \leq d$ for all $k = 0, 1, \ldots,$ as required. $\square$

From Theorems 3.1–3.3 we can deduce the following result.

**Theorem 3.4.** *If the assumptions of Theorem 3.3 are satisfied then there exist a subsequence of indices $k_1, k_2, \ldots$ and a positive constant $c$ such that*

$$f(x_{k_p}) - f(x^*) \leq c\, \alpha^{-k_p/n}, \quad p = 1, 2, \ldots.$$

*Moreover,*

$$\min_{1 \leq i \leq k} [f(x_i) - f(x^*)] \leq \frac{G\sqrt{k(\alpha^2 - 1)}\, d}{\sqrt{\alpha^{2k/n} - 1}},$$

*where $G = \max_{x \in S_d} \|g_f(x)\|$.*

*Proof.* Since the iterates remain in the bounded set $S_d$, the almost gradients are uniformly bounded and thus the conditions of Theorems 3.1 and 3.2 are satisfied.

Let $\{k_p\}_{p=1}^{\infty}$ be the sequence for which, according to Theorem 3.1, one has $\|\tilde{g}_{k_p}\| \leq \bar{c}\,\alpha^{-k_p/n}$. From (3.18) we obtain

$$N[f(x_{k_p}) - f(x^*)] \leq (g_f(x_{k_p}), x_{k_p} - x^*)$$
$$= (A_{k_p}^* \tilde{g}_{k_p}, x_{k_p} - x^*)$$
$$= (\tilde{g}_{k_p}, A_{k_p}(x_{k_p} - x^*))$$
$$\leq \|\tilde{g}_{k_p}\|\, d \leq \bar{c}\, d\, \alpha^{-k_p/n}.$$

Setting $\bar{c}\, d/N = c$ we get

$$f(x_{k_p}) - f(x^*) \leq \frac{\bar{c}\, d}{N}\, \alpha^{-k_p/n} = c\,\alpha^{-k_p/n}.$$

In a similar way, using Theorem 3.2 one obtains

$$\min_{1 \leq r \leq k}[f(x_r) - f(x^*)] \leq \min_{1 \leq r \leq k}\|\tilde{g}_r\|\,\frac{d}{N}\,,$$

whence

$$\min_{1 \leq r \leq k}[f(x_r) - f(x^*)] \leq \frac{G\sqrt{k(\alpha^2 - 1)}\, d}{N\sqrt{\alpha^{2k/n} - 1}}$$

This proves the result. $\quad\square$

One can extend Theorem 3.3 to the case of nonunique minimum points of the function $f$.

Suppose that the set $M^*$ of all minimum points of an almost differentiable function $f$ is bounded and the value of $f$ on $M^*$ is equal $f^*$.

Let us denote

$$\varrho_k(x_k) = \min_{x^* \in M^*}\|A_k(x_k - x^*)\| = \|A_k(x_k - x_k^*(x_k))\|;\ x_k^*(x_k) \in M^*.$$

**Theorem 3.5.** *Assume that for all $x$ and $x^*$ satisfying the condtions $\varrho_0(x) \leq d$, $x \notin M^*$ the following inequalities hold:*

$$N[f(x) - f^*] \leq (g_f(x), x - x^{**}) \leq M[f(x) - f^*],$$

*where $x^{**}$ is the point nearest to $x$ in $M^* \cap \{y: y = x + t(x^* - x), t \geq 0\}$. If the SDG algorithm is run with*

1) $\varrho_0(x_0) \leq d,$

2) $h_{k+1} = \dfrac{2MN}{M + N}\,\dfrac{f(x_k) - f^*}{\|g_{\varphi_k}(y_k)\|}\,,$

3) $1 < \alpha_{k+1} = \alpha \leq \dfrac{M + N}{M - N}\,,$

*then $\varrho_{k+1}(x_{k+1}) \leq d$ for $k = 0, 1, 2, \ldots$.*

*Proof.* We shall show that the sequence $\{\varrho_k(x_k)\}$ is nonincreasing:

$$\varrho_{k+1}^2(x_{k+1}) = \min_{y \in M^*} \| R_{\alpha_{k+1}}(\xi_{k+1}) A_k(x_{k+1} - y) \|^2$$

$$= \min_{y \in M^*} \| R_{\alpha_{k+1}}(\xi_{k+1}) A_k(x_k - y) - R_{\alpha_{k+1}}(\xi_{k+1}) h_{k+1} \xi_{k+1} \|^2$$

$$\leqq \| R_{\alpha_{k+1}}(\xi_{k+1}) A_k(x_k - x_k^*(x_k)) - \alpha_{k+1} h_{k+1} \xi_{k+1} \|^2.$$

Let $z_k = A_k(x_k - x_k^*(x_k))$. The remaining part of the proof is similar to that of Theorem 3.3. We obtain $\|z_{k+1}\| \leqq \|z_k\| \leqq \|z_0\| \leqq d, k = 0, 1, 2, \ldots,$ as required. $\square$

We shall use Theorem 3.5 for constructing a modification of the SDG algorithm in the case of unknown $f^*$, assuming tha the function $f$ is convex. Let us observe that in such a case one may always choose $N$ equal to unity, because $f(x) - f(x^*) \leqq (g_f(x), x - x^*)$ for all $x^* \in M^*$.

**Theorem 3.6.** *Let $f$ be a convex function possessing the following property: there exists a constant $M > 1$ such that if the function*

$$\varphi(\alpha) = f((1 - \alpha) x_1 + \alpha x_2)$$

*is strictly decreasing on $[0; 1]$ then*

$$(g_f(x_1), x_1 - x_2) \leqq M [f(x_1) - f(x_2)]. \tag{3.22}$$

*Furthermore, suppose that $\lim_{\|x\| \to \infty} f(x) = + \infty$. Then the SDG method with*

$$\alpha_{k+1} = \frac{M + 1}{M - 1},$$

$$h_{k+1} = \frac{2M}{M + 1} \frac{f(x_k) - \bar{f}}{\|\tilde{g}_k\|}$$

*has the following properties: if $\bar{f} \geqq f^*$ then the sequence $\{h_k\}$ is bounded and for any $\varepsilon > 0$ one can find $\bar{k}$ such that $f(x_{\bar{k}}) < \bar{f} + \varepsilon$ (we assume that if $f(x_k) \leqq \bar{f}$ at a certain step then the algorithm terminates); if $\bar{f} < f^*$ then the sequence $\{h_k\}$ is unbounded.*

*Proof.* Suppose that $\bar{f} \geqq f^*$ and consider the function

$$\bar{f}(x) = \begin{cases} f(x) & \text{if } f(x) \geqq \bar{f}, \\ \bar{f} & \text{if } f(x) < \bar{f}. \end{cases}$$

The assumptions of Theorem 3.5 are satisfied for this function, and so the sequence $\{\varrho_k(x_k)\}$ is bounded, which implies the convergence of function values by virtue of Theorem 3.4. The convexity of $\bar{f}(x)$ yields

$$\frac{f(x_k) - \bar{f}}{\|\tilde{g}_k\|} \leqq \varrho_k(x_k).$$

Hence $h_k \leqq \dfrac{2M}{M + 1} \varrho_k(x_k)$, i.e. $\{h_k\}$ is bounded.

If $\bar{f} < f^*$ then two cases are possible: (a) $\sup_k f(x_k) < \infty$; (b) $\sup_k f(x_k) = +\infty$.

In case (a) the conditions of Theorem 3.1 are satisfied, which implies the existence of a sequence $\{\|\tilde{g}_{k_p}\|\}_{p=0}^{\infty}$ converging to zero, and so

$$\sup_k h_k = \sup_k \frac{2M}{M+1} \frac{f(x_k) - \bar{f}}{\|\tilde{g}_k\|} \ge \sup_k \frac{f^* - \bar{f}}{\|\tilde{g}_k\|} = +\infty.$$

In case (b) we shall consider a ray originating at $x_k$ and passing through the set of minima $M^*$. Let $x_k^*$ be the first minimum point on this ray. By (3.22),

$$\frac{f(x_k) - f^*}{\|g_f(x_k)\|} \ge \frac{1}{M} \frac{(g_f(x_k), x_k - x_k^*)}{\|g_f(x_k)\|} = \frac{R_k}{M},$$

where $R_k$ is the distance from $x_k^*$ to the hyperplane supporting the level set $\{x : f(x) \le f(x_k)\}$ at $x_k$. Since the set $M^*$ is bounded and $\{f(x_k)\}$ is unbounded, $\sup_k R_k = +\infty$. Next,

$$\sup_k h_k = \sup_k \frac{2M}{M+1} \frac{f(x_k) - \bar{f}}{\|\tilde{g}_k\|} \ge \sup_k \frac{2M}{M+1} \frac{f(x_k) - \bar{f}}{\|g_f(x_k)\|}$$

$$\ge \frac{2}{M+1} \sup_k R_k = +\infty.$$

The proof is complete. $\quad\square$

We shall use the above result to construct an algorithm for minimizing a function $f$ with unknown $f^*$, assuming that $f$ satisfies the conditions of Theorem 3.6. The algorithm consists of a sequence of stages at which the SDG method of Theorem 3.6 is used.

At the beginning we choose $x_0^{(1)} \in E_n$, $h > 0$, $\varDelta_0 > 0$, $\tilde{f}_0 = f(x_0^{(1)})$. Assume that $r$ stages of the algorithm have been performed. At the beginning of the $(r+1)$-st stage we have

$$x_0^{(r+1)} \in E_n, \quad h > 0, \quad \varDelta_r > 0; \quad \tilde{f}_r = f(x_0^{(r+1)}),$$

where $x_0^{(r+1)}$ is the point with the smallest objective value $\tilde{f}_r$ obtained so far. We set

$$\varDelta_{r+1} = \begin{cases} \varDelta_r & \text{if } f(x_0^{(r+1)}) \le f(x_0^{(r)}) - \dfrac{\varDelta_r}{2}, \\[2ex] \dfrac{\varDelta_r}{2} & \text{if } f(x_0^{(r+1)}) > f(x_0^{(r)}) - \dfrac{\varDelta_r}{2}, \end{cases}$$

and $\bar{f}_{r+1} = f(x_0^{(r+1)}) - \varDelta_{r+1}$. Setting $x_0 = x_0^{(r+1)}$ and $\bar{f} = \bar{f}_{r+1}$ we use the SDG method of the form described in Theorem 3.6. After a finite number of steps, for some $k = \bar{k}_{r+1}$ one of the following cases must occur:

(a) $\quad f(x_{\bar{k}_{r+1}}) \le f(x_0^{(r+1)}) - \dfrac{\varDelta_{r+1}}{2};$

(b) $\quad h_{\bar{k}_{r+1}+1} > h.$

This completes the $(r+1)$-st stage. The "record" value $\tilde{f}_{r+1} = \min\limits_{0 \le k \le \bar{k}_{r+1}} f(x_k^{(r+1)})$ and the corresponding point $x_{k_*}^{(r+1)} = x_0^{(r+2)}$ are stored and the $(k+2)$-nd stage is initialized.

We shall prove that $\lim\limits_{r \to \infty} f(x_0^{(r)}) = f^*$. Since $\{f(x_0^{(r)})\}$ is nonincreasing and bounded from below, $\lim\limits_{r \to \infty} f(x_0^{(r)})$ exists. It is clear from the description of the algorithm and the proof of Theorem 3.6 that for a sufficiently small $\Delta_r$ case (b) may occur only if $\bar{f}_{r+1} < f^*$. On the other hand, once case (b) happens, case (a) may occur for only a finite number of subsequent stages. Therefore $\lim\limits_{r \to \infty} \Delta_r = 0$. Since for sufficiently small $\Delta_r$ (with $h$ fixed) case (b) implies that $f(x_0^{(r)}) - \Delta_r$ is a lower bound for $f^*$, we have

$$\lim\limits_{r \to \infty} f(x_0^{(r)}) = f^*,$$

as required.

Under certain circumstances, e.g. when solving a system of nonlinear equations (see Sect. 3.5), we know $f^*$, but difficulties may arise in the estimation of $M$ and $N$. If the constants $M$ and $N$ are chosen improperly, the $(M, N)$-*algorithm with space dilation* described in Theorem 3.3 may converge too slowly or even diverge. Therefore it is important to have tests for detecting such "pathological" situations. The simplest test is based on the analysis of the behavior of the sequence $\{h_k\}$.

**Theorem 3.7.** *If the parameters $M$ and $N$ in the $(M, N)$-algorithm are chosen correctly then the sequence $\{h_k\}$ is bounded. If the values of these parameters are wrong and the algorithm does not converge, then either $\{h_k\}$ or $\{f(x_k)\}$ is unbounded.*

One can prove the above theorem similarly as Theorem 3.6.

To sum up, if $h_k$ or $f(x_k)$ exceed sufficiently large bounds $h^{\max}$ or $f^{\max}$, one should increase the constant $M$, decrease $N$ and restart the SDG algorithm of Theorem 3.3 either from the initial point or from the best point obtained so far.

## 3.5 Application of the Subgradient Method with Space Dilation to the Solution of Systems of Nonlinear Equations

The rules for determining coefficients of space dilation and stepsizes described in Theorem 3.3 are readily applicable when one solves a system of nonlinear equations

$$\psi_i(x) = 0, \quad i = 1, \ldots, n, \quad x = \{t_1, \ldots, t_n\} \tag{3.23}$$

by minimizing the nonsmooth function

$$f(x) = \max_{1 \le i \le n} |\psi_i(x)|. \tag{3.24}$$

If the functions $\psi_i$, $i = 1, \ldots, n$, are almost differentiable, then $f$ is also almost differentiable (see Theorem 1.17) and can be minimized by the SDG method with the stepsize and dilation rules of Theorem 3.3, since one has $f^* = \min_{x \in E_n} f(x) = 0$, whenever system (3.23) is consistent.

Let us discuss the problem of choosing the values of the parameters $M$ and $N$ in inequality (3.18) for $f$ given by (3.24). If we restrict our considerations to functions $\psi_i$ which are continuously differentiable almost everywhere, then the following situation may be considered typical (regular): $f^* = 0$, the functions $\psi_i$ are continuously differentiable in a neighborhood of the minimum of $f$ and the Jacobian $J(x) = \{\partial \psi_i / \partial t_j\}$ is nonsingular at the minimum.

**Theorem 3.8.** *In the regular case for any $\delta > 0$ one can find a neighborhood of the optimal point $x^*$ such that*

$$(1 - \delta) f(x) \le (g_f(x), x - x^*) \le (1 + \delta) f(x)$$

*for all $x$ in that neighborhood.*

*Proof.* We shall start by proving that the derivatives of $f$ at $x^*$ are positive in all directions. It suffices to show that they are all different from zero. Were it not true, one could find a nonzero direction $\xi$ orthogonal to $g_{\psi_i}(x^*)$, $i = 1, \ldots, n$, which would contradict the linear independence of the system of vectors $\{g_{\psi_i}(x^*)\}_{i=1}^n$.

Since the directional derivatives $f'_\xi(x^*)$ at $x^*$ depend continuously on $\xi$ and the unit sphere $\{\xi \in E_n : \|\xi\| = 1\}$ is compact, one can find $\delta' > 0$ such that $f'_\xi(x^*) \ge \delta'$ for every $\xi$ with $\|\xi\| = 1$.

For each direction $\xi$, $\|\xi\| = 1$, the directional derivatives of the functions $\psi_i$ are uniformly continuous in a certain neighborhood of the point $x^*$. Consequently, there exists a ball around $x^*$ in which

$$|\psi'_{i\xi}(x) - \psi'_{i\xi}(x^*)| \le \frac{\delta' \delta}{3}.$$

Take any point $x$ of that ball and consider the directional derivative of $f$ in the direction $\mu(x) = (x - x^*)/\|x - x^*\|$. The derivative satisfies the relations

$$f'_{\mu(x)}(x) - \frac{\delta \delta'}{3} \le f'_{\mu(x)}(x^*) \le f'_{\mu(x)}(x) + \delta \delta'. \tag{3.25}$$

Indeed,

$$f'_{\mu(x)}(x) = \max_{i \in I(x)} [\psi'_{i\mu(x)}(x) \operatorname{sign} \psi_i(x)],$$

where $I(x)$ is the set of indices for which

$$f(x) = |\psi_i(x)|.$$

On the other hand,

$$f'_{\mu(x)}(x^*) = |\psi'_{i_0\mu(x)}(x^*)| = \max_{1 \leq i \leq n} |\psi'_{i\mu(x)}(x^*)|.$$

If $i_0 \in I(x)$ then (3.25) is obvious. If $i_0 \notin I(x)$ then in the segment $\{\alpha x^* + (1 - \alpha) x, 0 \leq \alpha \leq 1\}$ one can find a point $x_0$ at which

$$\psi'_{\bar{i}\mu(x)}(x_0) \, \text{sign} \, \psi_{\bar{i}}(x_0) \geq \psi'_{i_0\mu(x)}(x_0) \, \text{sign} \, \psi_{i_0}(x_0) \geq |\psi'_{i_0\mu(x)}(x^*)| - \frac{\delta\,\delta'}{3}$$

for a certain $\bar{i} \in I(x)$, whence

$$|\psi'_{\bar{i}\mu(x)}(x^*)| \geq |\psi'_{i_0\mu(x)}(x^*)| - \frac{2\,\delta\,\delta'}{3},$$

$$f'_{\mu(x)}(x) \geq |\psi'_{\bar{i}\mu(x)}(x^*)| - \frac{\delta\,\delta'}{3} \geq f'_{\mu(x)}(x^*) - \delta\,\delta'.$$

The proof is complete. $\quad\square$

It follows from the above theorem that in the regular case in a sufficiently small neighborhood of $x^*$ we can use large $\alpha$ and $\gamma = 2MN/(M + N)$ close to unity, thus obtaining rather fast convergence. Indeed, if $N = 1 - \delta$ and $M = 1 + \delta$ then $\alpha$ may be chosen equal to $(M + N)/M - N = 1/\delta$ and $\gamma = 2MN/(M + N) = 1 - \delta^2$. If $\delta \to 0$ then $\gamma \to 1$ and $\alpha \to \infty$. But if the starting point $x_0$ is far from the solution then it is difficult to estimate the parameters $M$ and $N$. In that case a special procedure is needed for correcting initial estimates of $M$ and $N$ in the course of computation. The case of improper values of $M$ and $M$ can, according to the results of Sect. 3.4, be detected by analysing the behaviour of the parameters $\{h_k\}$. If $h_k$ exceeds a prespecified bound (e.g. equal to an upper estimate of the distance from $x_0$ to the minimum), then we should increase $M$, decrease $N$, recompute $\gamma$ and $\alpha$, and continue the calculations from the point at which $f$ attained the smallest value at previous iterations.

It should be noted that reducing the system of nonlinear equations (3.23) to a minimax problem is much more efficient than other transformations to optimization problems (e.g. the minimization of the sum of either the squares or the absolute values of the residuals), since the SDG method requires an evaluation of the gradient (almost-gradient) of only one of the functions $\psi_i$ at each iteration, whereas the other methods require the evaluation of all the gradients. Moreover, an extreme variant of the SDG algorithm with $\beta = 0$ ($\alpha = +\infty$) is in fact a gradient orthogonalization method. When applied to a system of nonlinear equations, it converges at a quadratic rate under certain additional conditions. Let us consider this method in detail. Assume that we have a system of nonlinear equations

$$\psi_i(x) = 0, \quad i = 1, 2, \ldots, n, \quad x \in E_n, \tag{3.26}$$

where $\psi_i(x) = \psi_i(t_1, t_2, \ldots, t_n)$ are continuously differentiable functions. Let us define an algorithm for solving (3.26) which consists of a sequence of stages, each comprising $n$ successive steps. Let $x_0^{(r)}$ be the starting point for the $r$-th stage.

*Step 1.* Evaluate $\psi_1(x_0^{(r)})$ and the gradient $g_{\psi_1}(x_0^{(r)})$ and compute

$$x_1^{(r)} = x_0^{(r)} - \frac{\psi_1(x_0^{(r)})}{\|g_{\psi_1}(x_0^{(r)})\|} \, g_{\psi_1}(x_0^{(r)}).  \tag{3.27a}$$

*Step $k+1$ $(1 \le k < n)$.* Evaluate $\psi_{k+1}(x_k^{(r)})$ and $g_{\psi_{k+1}}(x_k^{(r)})$, compute the vector $\varphi_{k+1}^{(r)}$ by projecting the vector $g_{\psi_{k+1}}(x_k^{(r)})$ on the subspace orthogonal to the vectors $g_{\psi_1}(x_0^{(r)}), \ldots, g_{\psi_k}(x_{k-1}^{(r)})$, and find

$$x_{k+1}^{(r)} = x_k^{(r)} - \frac{\psi_{k+1}(x_k^{(r)})}{\|\varphi_{k+1}^{(r)}\|^2} \, \varphi_{k+1}^{(r)}.  \tag{3.27b}$$

The point $x_n^{(r)}$ thus obtained is the initial approximation for the next stage: $x_0^{(r+1)} = x_n^{(r)}$.

We shall show that if the initial guess $x_0^{(1)}$ is sufficiently close to the solution $x^*$ of the system and some other conditions are satisfied, then the sequence $x_0^{(1)}, \ldots, x_0^{(r)}, \ldots$, converges to the point $x^*$ at a quadratic rate. Without loss of generality we assume that $x^* = 0$.

**Theorem 3.9.** *Let $x^* = 0$ be the solution to the system of equations (3.26). Let the functions $\psi_i$, $i = 1, 2, \ldots, n$, be continuously differentiable and let their gradients satisfy the Lipschitz condition in a certain ball $S_\delta = \{x: \|x\| < \delta\}$, i.e. there is a constant $L$ such that*

$$\|g_{\psi_i}(x') - g_{\psi_i}(x'')\| \le L\|x' - x''\| \quad \text{for all } x', x'' \in S_\delta.  \tag{3.28}$$

*Additionally, let the vectors $\{g_{\psi_i}(0)\}_{i=1}^n$ be linearly independent, so that the determinant of the Jacobian*

$$\det J(x) = \det \left\{\frac{\partial \psi_i}{\partial t_j}\right\}_{i,j=1}^n$$

*is nonzero at the point $x^* = 0$. Then one can find $\varepsilon > 0$ such that if $\|x_0^{(r)}\| \le \varepsilon$ then*

$$\|x_n^{(r)}\| = \|x_0^{(r+1)}\| \le c\|x_0^{(r)}\|^2,$$

*where $c$ is a positive constant.*

*Proof.* Consider the function

$$D(x_0, \ldots, x_{n-1}) = \det \left\{\frac{\partial \psi_i(x_{i-1})}{\partial t_j}\right\}_{i,j=1}^n.$$

This function is continuous and $D(0, 0, \ldots, 0) = \det J(0) \ne 0$. Consequently, there exist a neighborhood $S_{\delta'}$ of $x^*$ (we take $\delta' \le \delta$) and a constant $a > 0$ such that

$$|D(x_0, \ldots, x_{n-1})| \ge a \quad \text{for all } x_i \in S_{\delta'}, i = 0, 1, \ldots, n-1.$$

Let $x_0^{(r)}, x_1^{(r)}, \ldots, x_{n-1}^{(r)} \in S_\delta$. Then one can find $b > 0$ such that

$$\| \varphi_{k+1}^{(r)} \| > b \quad \text{for } k = 0, 1, \ldots, n - 1. \tag{3.29}$$

Indeed, the system of vectors $\{\varphi_{k+1}^{(r)}\}$, $k = 0, 1, \ldots, n - 1$, results from the orthogonalization of the system of linearly independent vectors $\{g_{\psi_{k+1}}(x_k^{(r)})\}$. The Grammian of the system $\{\varphi_{k+1}^{(r)}\}$ is equal to the Grammian of the system $\{g_{\psi_{k+1}}(x_k^{(r)})\}$ which in turn is equal to

$$D^2(x_0^{(r)}, \ldots, x_{n-1}^{(r)}) \geqq a^2.$$

We obtain the inequality

$$\prod_{k=0}^{n-1} \| \varphi_{k+1}^{(r)} \| \geqq a. \tag{3.30}$$

Since the norms $\| \varphi_{k+1}^{(r)} \|$, $k = 0, 1, \ldots, n - 1$, are bounded from above, it follows from inequality (3.30) that they are bounded from below by a certain positive constant, which we denote by $b$. Next,

$$\begin{aligned}
\psi_{k+1}(x_{k+1}^{(r)}) &= \psi_{k+1}\left[ x_k^{(r)} - \frac{\psi_{k+1}(x_k^{(r)})}{\| \varphi_{k+1}^{(r)} \|^2} \varphi_{k+1}^{(r)} \right] \\
&= \psi_{k+1}(x_k^{(r)}) - \left( g_{\psi_{k+1}}^c, \frac{\varphi_{k+1}^{(r)}}{\| \varphi_{k+1}^{(r)} \|} \right) \frac{\psi_{k+1}(x_k^{(r)})}{\| \varphi_{k+1}^{(r)} \|},
\end{aligned}$$
$$k = 0, 1, \ldots, n - 1, \tag{3.31}$$

where $g_{\psi_{k+1}}^c$ is the value of the gradient at a certain point on the segment joining $x_k^{(r)}$ and $x_{k+1}^{(r)}$. By (3.28),

$$g_{\psi_{k+1}}^c = g_{\psi_{k+1}}(x_k^{(r)}) + \Delta g_{k+1}, \tag{3.32}$$

where

$$\| \Delta g_{k+1} \| \leqq \frac{L \, \psi_{k+1}(x_k^{(r)})}{\| \varphi_{k+1}^{(r)} \|}.$$

Substituting $g_{\psi_{k+1}}^c$ in (3.31) by the right side of (3.32) and taking into account that

$$(g_{\psi_{k+1}}(x_k^{(r)}), \varphi_{k+1}^{(r)}) = \| \varphi_{k+1}^{(r)} \|^2,$$

we obtain

$$| \psi_{k+1}(x_{k+1}^{(r)}) | \leqq \frac{L \, \psi_{k+1}^2(x_k^{(r)})}{\| \varphi_{k+1}^{(r)} \|^2} \leqq \frac{L}{b^2} \, \psi_{k+1}^2(x_k^{(r)}),$$
$$\text{for } k = 0, 1, \ldots, n - 1. \tag{3.33}$$

In the remaining part of the proof we shall use the following lemma.

**Lemma 3.1.** *Let $x_1$ and $x_2$ be two points, let $\xi$ be a vector such that $\| \xi \| = 1$ and $(g_{\psi_i}(x_1), \xi) = 0$ for some $i$, $1 \leqq i \leqq n$, and let $h$ be a positive constant. If*

$x_1, x_2, x_2 + h\,\xi \in S_\delta$ then

$$|\psi_i(x_2 + h\,\xi) - \psi_i(x_2)| \leq L\,h\,(h + d),$$

where $d = \|x_1 - x_2\|$.

*Proof.* By virtue of the Lagrange theorem,

$$|\psi_i(x_2 + h\,\xi) - \psi_i(x_2)| \leq h\,|(g_{\psi_i}(x_2 + \alpha\,h\,\xi), \xi)|,$$

where $0 \leq \alpha \leq 1$, and

$$|(g_{\psi_i}(x_2 + \alpha\,h\,\xi), \xi)| = |(g_{\psi_i}(x_2 + \alpha\,h\,\xi) - g_{\psi_i}(x_1), \xi)|$$

$$\leq \|g_{\psi_i}(x_2 + \alpha\,h\,\xi) - g_{\psi_i}(x_1)\| \leq L\,(h + d).$$

Hence

$$|\psi_i(x_2 + h\,\xi) - \psi_i(x_2)| \leq L\,h\,(h + d),$$

as required.

Let us continue the proof of Theorem 3.9. Assume that the points $x_0^{(r)}, \ldots, x_n^{(r)}$ lie in the sphere $S_\mu = \{x : \|x\| \leq \mu\}$ with $\mu \leq \delta'$. Then

$$|\psi_i(x_k^{(r)})| \leq \mu\,\|g_{\psi_i}(0)\| + \frac{L\,\mu^2}{2}, \qquad i = 1, \ldots, n; \quad k = 1, \ldots, n. \tag{3.34}$$

Inequalities (3.33) and (3.34) yield

$$|\psi_{k+1}(x_{k+1}^{(r)})| \leq \frac{L\,\mu^2}{b^2}\left[\|g_{\psi_{k+1}}(0)\| + \frac{L\,\mu}{2}\right]^2.$$

From Lemma 3.1 we obtain the inequality

$$|\psi_{k+1}(x_n^{(r)})| \leq |\psi_{k+1}(x_{k+1}^{(r)})| + \sum_{i=1}^{n-k-1} |\psi_{k+1}(x_{k+i+1}^{(r)}) - \psi_{k+1}(x_{k+i}^{(r)})|$$

$$\leq \frac{L\,\mu^2}{b^2}\left[\|g_{\psi_{k+1}}(0)\| + \frac{L\,\mu}{2}\right]^2 + \sum_{i=1}^{n-k-1} L \cdot 2\,\mu \cdot 4\,\mu$$

$$= \bar{c}\,\mu^2 + o(\mu^2), \tag{3.35}$$

where $\bar{c}$ is a certain constant. The expression for $\bar{c}$ may be easily derived from (3.35).

The proof will be complete if we verify the following statement: there exist a number $\varepsilon_0$ and two positie constants $c_1$ and $c_2$ such that if $\|x_0^{(r)}\| \leq \varepsilon_0$ then

$$(1) \quad \max_{1 \leq k \leq n} \|x_k^{(r)}\| \leq c_1\,\|x_0^{(r)}\|, \tag{3.36}$$

$$(2) \quad f(x_n^{(r)}) = \max_{0 \leq k \leq n-1} \|\psi_{k+1}(x_n^{(r)})\| \geq c_2\,\|x_n^{(r)}\|. \tag{3.37}$$

The first assertion may be verified by estimating directly the norms of the shifts $\| x_{k+1}^{(r)} - x_k^{(r)} \|$, $k = 0, 1, \ldots, n - 1$, with the application of formulas (3.27) and the estimates (3.29), (3.34). The second assertion follows from the first one and the fact that in a certain neighborhood of the solution the function $f$ has directional derivatives bounded from below by a positive constant:

$$f'_{x/\|x\|}(x) \geqq d > 0.$$

Let $\| x_0^{(r)} \| = \varepsilon \leqq \max(\varepsilon_0, \delta'/c_1)$. Then $(3.35)-(3.37)$ yield

$$\max_{0 \leqq k \leqq n-1} | \psi_{k+1}(x_n^{(r)}) | \leqq \bar{c} \, c_1^2 \, \varepsilon^2 + o(c_1^2 \, \varepsilon^2) \leqq k_1 \, c_1^2 \, \varepsilon^2,$$

whence

$$\| x_n^{(r)} \| \leqq \frac{k_1 \, c_1^2}{c_2} \varepsilon^2 = c \, \| x_0^{(r)} \|^2,$$

where $c = k_1 \, c_1^2/c_2$. This completes the proof.   $\square$

## 3.6 A Minimization Method Using the Operation of Space Dilation in the Direction of the Difference of Two Successive Almost-Gradients

The problem of stepsize selection is crucial for developing various versions of the SDG method. It can be solved easily if the minimum value of the function is known in advance. In general, however, one has to complicate the algorithm.

We shall describe another class of algorithms, called *r-algorithms*, which are based on the application of the operation of space dilation along the difference of two successive almost gradients. Contrary to the SDG methods, such algorithms may use stepsize rules similar to that of the method of steepest descent. The algorithms thus obtained decrease the function values at every or almost every iteration.

The results of the present section are based on the works of the author and N. G. Zhurbenko [82, 95]. Applications to minimax problems were studied in detail by L. P. Shabashova [72, 93].

Let us now pass on to the description of *r*-algorithms.

Let $f$ be an almost differentiable function defined on the $n$-dimensional Euclidean space $E_n$ and satisfying the following condition:

$$\lim_{\|x\| \to +\infty} f(x) = + \infty.$$

We denote by $g_f(x)$ an almost-gradient of $f$ at $x$. Let us consider the following iterative algorithm for minimizing $f$.

At the first iteration, for a given starting point $x_0$ we compute $g_f(x_0)$, choose $h_1 > 0$ and find $x_1 = x_0 - h_1 g_f(x_0)$. We set $\tilde{g}_1 = g_f(x_0)$, $B = I$ (the $n \times n$ unit matrix), and we proceed to the second iteration.

Assume that after $k$ iterations ($k = 1, 2, \ldots$) we have vectors $x_k$, $\tilde{g}_k \in E_n$ and a matrix $B_k$. We shall describe the $(k + 1)$-st iteration of the algorithm.

The following quantities are calculated:

(1) $g_f(x_k)$ – the almost-gradient of $f$ at $x_k$;

(2) $g_k^* = B_k^* g_f(x_k)$ – the almost gradient of the function $\varphi_k(y) = f(B_k y)$ at the point $y = A_k x_k$, where $A_k = B_k^{-1}$;

(3) $\qquad r_k = g_k^* - \tilde{g}_k$ $\hfill$ (3.38)

– the difference of two gradients of the function $\varphi_k(y)$ computed at the points $y_k = A_k x_k$ and $\tilde{y}_k = A_k x_{k-1}$ ($r_k$ may be computed also in a different way:

$$r_k = B_k^* [g_f(x_k) - g_f(x_{k-1})], \quad k = 1, 2, \ldots,$$

but the application of formula (3.38) reduces the computational effort);

(4) $\xi_{k+1} = r_k / \| r_k \|$ – this normalization of $r_k$ will be used for the next space dilation along $r_k$;

(5) $\beta_{k+1}$ – the inverse of the coefficient of space dilation;

(6) $B_{k+1} = B_k R_{\beta_{k+1}}(\xi_{k+1})$ – the inverse of the new space transformation matrix $A_{k+1} = R_{\alpha_{k+1}}(\xi_{k+1}) A_k$ ($R_\alpha(\xi)$ is the operator of space dilation in the direction $\xi$ with the coefficient $\alpha = 1/\beta$; $\| \xi \| = 1$);

(7) $h_{k+1}$ – the stepsize;

(8) $\tilde{g}_{k+1} = R_{\beta_{k+1}}(\xi_{k+1}) g_k^* = R_{\beta_{k+1}}(\xi_{k+1}) B_k^* g_f(x_k)$

$\qquad = (B_k R_{\beta_{k+1}}(\xi_{k+1}))^* g_f(x_k) = B_{k+1}^* g_f(x_k)$

– the value of the gradient of the function $\varphi_{k+1}(y) = f(B_{k+1} y)$ at the point $\tilde{y}_{k+1} = A_{k+1} x_k$;

(9) $x_{k+1} = x_k - h_{k+1} B_{k+1} \tilde{g}_{k+1}$ $\hfill$ (3.39)

(observe that applying the operator $A_{k+1}$ to both sides of formula (3.39) we obtain

$$y_{k+1} = A_{k+1} x_{k+1} = A_{k+1} x_k - h_{k+1} \tilde{g}_{k+1} = \tilde{y}_{k+1} - h_{k+1} \tilde{g}_{k+1},$$

which is in fact a step of the subgradient method for the function $\varphi_{k+1}(y)$);

(10) if a stopping test is satisfied then the computations terminate; otherwise $x_{k+1}$, $\tilde{g}_{k+1}$ and $B_{k+1}$ are stored and the $(k + 2)$-nd iteration is executed.

From the above-described class of gradient-type methods with a variable metric we obtain definite algorithms by specifying rules for determining sequences $\{h_{k+1}\}$ and $\{\beta_{k+1}\}$ and the stopping criterion.

The $r$-algorithms may be further modified by using the so-called resetting operation, in which periodically, after a specified number of iterations, the matrix $B_k$ is "reset", i.e. replaced with the unit matrix. The analysis of convergence of descent algorithms with resets is relatively easy, because it actually reduces to the analysis of convergence of the method of steepest descent (without the variable metric), which may be found in many works (see, e.g., [66]). In particular, if $f$ is continuously differentiable and $h_{k+1}$ is chosen by exact directional minimization, then the set of accumulation points of the sequence $\{x_k\}$, generated by the $r$-algorithm with resetting consists of stationary points for $f$, which have the same value of $f$.

The proof of this assertion does not essentially differ from the proof of a similar result for the method of steepest descent. But if the resetting is not used, the proof of the convergence of $r$-algorithm requires more effort.

Let us consider in a more detailed way an extreme variant of the $r$-algorithm with resetting, in which $\beta_k, k = 1, 2, \ldots$, are chosen equal to zero and $h_{k+1}$ is determined by minimizing $f(x_k - h\, B_{k+1}\, \tilde{g}_{k+1})$. As one may easily observe, one has then either $\tilde{g}_{k*+1} = 0$ for some $k* < n$, or $B_{n+1} = 0$. Indeed, $R_0(\eta)$ is the operator of projection on the subspace orthogonal to the vector $\eta$. The product of such operators, $\prod_{i=1}^{k} R_0(\eta_i)$, does not depend on the order of factors, is a selfadjoint operator, and realizes the orthogonal projection on the supspace being the orthogonal complement of the linear span of the vectors $\eta_i, i = 1, \ldots, k$.

We have the following equations:

$$r_1 \;\; = g_f(x_1) - g_f(x_0),$$

$$\cdots \cdots \cdots \cdots \cdots$$

$$r_k \;\; = R_0\left(\frac{r_{k-1}}{\|r_{k-1}\|}\right) R_0\left(\frac{r_{k-2}}{\|r_{k-2}\|}\right) \ldots R_0\left(\frac{r_1}{\|r_1\|}\right) \Delta g_k;$$

$$\Delta g_k = g_f(x_k) - g_f(x_{k-1}), \quad k > 1.$$

We see that the vectors $r_1, \ldots, r_k$ result from the successive orthogonalization of the vectors $g_f(x_1) - g_f(x_0), g_f(x_2) - g_f(x_1), \ldots, g_f(x_k) - g_f(x_{k-1})$, $k = 1, 2, \ldots$. If $r_1, \ldots, r_n$ differ from the null vector, then $B_{n+1} = 0$, because the operator $B_{n+1}$ realizes the projection on the orthogonal complement of the whole space. If $r_k = 0$ for some $k \leq n$, then

$$B_k^* [g_f(x_k) - g_f(x_{k-1})] = 0,$$

which is possible only if $B_k^* g_f(x_{k-1}) = \tilde{g}_k = 0, x_k = x_{k-1}$.

Therefore, in the variant of the $r$-algorithm under consideration, one has to apply the resetting when $B_k^* g_f(x_{k-1}) = 0$. This variant may be regarded as a modification of the conjugate gradient method. It is easy to prove that for a nonnegative definite quadratic form the solution is found in no more than $n$ iterations. For any sufficiently smooth function the $n$-step quadratic rate of convergence may be proved.

**Theorem 3.10.** *Let $f$ be a function defined on $E_n$ and twice continuously differentiable in a certain neighborhood $S$ of the optimal point $x^*$, and let the* *matrix of the second derivatives (the Hessian) $H(x)$ be Lipschitz in that neighborhood:*

$$\| H(x) - H(x') \| \leq L \| x - x' \| \qquad \text{for all } x, x' \in S. \tag{3.40}$$

*Furthermore, suppose that $H(x^*)$ is a positive definite matrix. Then there exists a neighborhood $S'$ of $x^*, S' \subset S$, such that if $x_0 \in S'$ then one can find a constant $c$ such that*

$$\| x_n - x^* \| \leq c \| x_0 - x^* \|^2,$$

*where $x_n$ is the point obtained after n iterations of the algorithm described above (if $g_{\bar{k}} = 0$ for some $\bar{k} < n$, then we set $x_n = x_{\bar{k}}$).*

*Proof.* With no loss of generality we assume that $x^* = 0$ and $f(0) = 0$. We shall prove the theorem by induction with respect to the dimension of the space $E_n$. For $n = 1$ the assertion is trivial. Suppose that the theorem is true for $n = p$. We shall prove it for $n = p + 1$ ($p \geq 1$). Let us denote the largest eigenvalue of the operator $H(0)$ by $M_0$ and the smallest one by $m_0$. By virtue of the positive definiteness of $H(0)$, $m_0 > 0$. It follows from the continuity of the Hessian $H(x)$ that there exist a neighborhood $S_\delta = \{x: \|x\| < \delta\}$ of $x^*$ and positive numbers $M$ and $m$, such that for $x \in S_\delta$ the largest eigenvalue of the operator $H(x)$ does not exceed $M$ and the smallest one is not less than $m$ ($m > 0$). Let $e$ be a vector with $\| e \| = 1$. For $x_0 = \varepsilon e, 0 < \varepsilon < \delta$, we have

$$g_f(x_0) = \int_0^\varepsilon H(\mu e) \, e \, d\mu = \int_0^\varepsilon [H(0) + G(\mu e)] \, e \, d\mu.$$

By (3.40) $\| G(\mu e) \| \leq \mu L$, whence

$$g_f(x_0) = \varepsilon [H(0) \, e + r(x_0)], \tag{3.41}$$

where

$$\| r(x_0) \| \leq \frac{1}{\varepsilon} \int_0^\varepsilon \mu L \, d\mu = \frac{L \varepsilon}{2}. \tag{3.42}$$

Let us determine $x_1$ as in our algorithm:

$$x_1 = x_0 - h^* \frac{g_f(x_0)}{\| g_f(x_0) \|},$$

where $h^*$ is the smallest positive root of the equation

$$\left( g_f(x_0), g_f\left( x_0 - h^* \frac{g_f(x_0)}{\| g_f(x_0) \|} \right) \right) = 0.$$

It is easy to observe that for a sufficiently small $\varepsilon$ one has

$$h^* \leq \frac{\| g_f(x_0) \|}{m} \leq \frac{M \varepsilon}{m}.$$

Indeed, let

$$x(h) = x_0 - h \frac{g_f(x_0)}{\| g_f(x_0) \|}, \qquad \varepsilon < \frac{\delta}{\dfrac{M}{m} + 1}.$$

Consider the function $\varphi(h) = f(x(h))$. For $0 \le h \le M \varepsilon/m$ one has $x(h) \in S_\delta$. Then the properties of the Hessian in $S_\delta$ imply that

$$\frac{d^2\varphi(h)}{dh^2} \ge m,$$

whence

$$\frac{d\varphi}{dh} \le \| g_f(x_0) \| - mh \le M\varepsilon - mh.$$

Thus $h^* \le M \varepsilon/m$, as required.

Next,

$$g_f(x_1) = g_f\left( x_0 - h^* \frac{g_f(x_0)}{\| g_f(x_0) \|} \right)$$

$$= g_f(x_0) - \int_0^{h^*} H\left( x_0 - \mu \frac{g_f(x_0)}{\| g_f(x_0) \|} \right) \frac{g_f(x_0)}{\| g_f(x_0) \|} \, d\mu$$

$$= g_f(x_0) - \int_0^{h^*} \left[ H(x_0) + G_1\left( \mu \frac{g_f(x_0)}{\| g_f(x_0) \|} \right) \right] \frac{g_f(x_0)}{\| g_f(x_0) \|} \, d\mu,$$

where $\left\| G_1\left( \mu \dfrac{g_f(x_0)}{\| g_f(x_0) \|} \right) \right\| \le L \mu.$ Hence

$$g_f(x_1) = g_f\left( x_0 - h^* \frac{g_f(x_0)}{\| g_f(x_0) \|} \right)$$

$$= g_f(x_0) - h^* H(x_0) \frac{g_f(x_0)}{\| g_f(x_0) \|} + h^* r_1(x_0) \tag{3.43}$$

with

$$\| r_1(x_0) \| \le \frac{L h^*}{2} \le \frac{LM \varepsilon}{2 m}. \tag{3.44}$$

From (3.41)–(3.43) we obtain

$$(g_f(x_0), g_f(x_1)) = \varepsilon^2 \Big[ (H(0) e + r(x_0), H(0) e + r(x_0))$$

$$- \frac{h^*}{\| g_f(x_0) \|} ((H(0) + G(x_0)) (H(0) e + r(x_0)), H(0) e + r(x_0)) \Big]$$

$$+ \varepsilon h^* (H(0) e + r(x_0), r_1(x_0))$$

$$= \varepsilon^2 \Big[ (H(0) \, e, H(0) \, e) - \frac{h^*}{\| g_f(x_0) \|} (H^2(0) \, e, H(0) \, e) \Big] + r_2(x_0)$$

where $| r_2(x_0) | \le d\varepsilon^3$ for all sufficiently small $\varepsilon$ and some positive constant $d$.

Setting $(g_f(x_0), g_f(x_1)) = 0$ we obtain

$$1 - \frac{d\varepsilon}{\|H(0)\,e\|^2}\,q \leq h^* \leq 1 + \frac{d\varepsilon}{\|H(0)\,e\|^2}\,q,$$

where

$$q = \frac{\|H(0)\,e\|^2\,g_f(x_0)}{(H^2(0)\,e,\,H(0)\,e)}.$$

We shall now show that the distance from the origin to the hyperplane that contains $x_0$ and has the normal vector $\varDelta = g_f(x_0) - g_f(x_1)$ is of order $\varepsilon^2$. Since this hyperplane is defined by $(x - x_1, g_f(x_0) - g_f(x_1)) = 0$, its distance to the origin may be computed as follows:

$$S = \frac{|g_f(x_0) - g_f(x_1),\, x_1)|}{\|g_f(x_0) - g_f(x_1)\|}.$$

We also have

$$x_1 = x_0 - h^* \frac{g_f(x_0)}{\|g_f(x_0)\|} = x_0 \frac{\|H(0)\,e\|^2\,g_f(x_0)}{(H^2(0)\,e,\,H(0)\,e)} - \frac{v\,q\,g_f(x_0)}{\|g_f(x_0)\|}, \tag{3.45}$$

where

$$-\frac{d\varepsilon}{\|H(0)\,e\|^2} \leq v \leq \frac{d\varepsilon}{\|H(0)\,e\|^2}. \tag{3.46}$$

Next, from $(3.43) - (3.46)$ we obtain

$$
\begin{aligned}
(g_f(x_0) - g_f(x_1), x_1) = {} & \frac{h^*\,\varepsilon^2}{\|g_f(x_0)\|} \Bigg( (H(0) + G(x_0))\,(H(0)\,e + r(x_0)), \\
& e - \frac{\|H(0)\,e\|^2\,(H(0)\,e + r(x_0))}{(H^2(0)\,e,\,H(0)\,e)} \Bigg) \\
& - \frac{1}{\|g_f(x_0)\|}\,(g_f(x_0),\, g_f(x_0) - g_f(x_1)) \\
& - h^*\,(r_1(x_0),\, x_1) \\
= {} & \frac{\varepsilon^2\,h^*}{\|g_f(x_0)\|}\,[(H^2(0)\,e,\,e) - (H(0)\,e,\,H(0)\,e)] + r_3(x_0),
\end{aligned}
$$

where $|r_3(x_0)| \leq c_1\,\varepsilon^2$ with some positive constant $c_1$ (which can be easily computed from the estimates (3.42), (3.44), (3.46)). Since $(g_f(x_0), g_f(x_1)) = 0$,

$$\|g_f(x_1) - g_f(x_0)\| \geq \|g_f(x_0)\| \geq m\,\varepsilon.$$

Hence

$$S \leq \frac{c_1}{m}\,\varepsilon^2 = c_2\,\varepsilon^2 \tag{3.47}$$

with $c_2 = c_1/m$. Consider the function $\psi$ which is the restriction of $f$ to the $p$-dimensional hyperplane

$$L(x_0) = \{x : (g_f(x_1) - g_f(x_0), x - x_1) = 0\}.$$

We shall prove that $\min \psi(x) = \min\limits_{x \in L(x_0)} f(x)$ is attained at a point $x_1^*$ for which

$$\|x_1^*\| \leq c_3 \, \varepsilon^2,$$

where $c_3$ is a positive constant independent of $\varepsilon$ for sufficiently small $\varepsilon$.

Define $\bar{x}_1$ as the point of $L(x_0)$ that is nearest to the origin. From (3.47) we obtain $S = \|\bar{x}_1\| \leq c_2 \, \varepsilon^2$, whence

$$\min\limits_{x \in L(x_0)} f(x) \leq f(\bar{x}_1) \leq \frac{M \, \varepsilon^4 \, c_2^2}{2} = c_4 \, \varepsilon^4.$$

On the other hand

$$f(x_1^*) \geq \frac{m \, \|x_1^*\|^2}{2} \, .$$

Consequently,

$$\frac{m \, \|x_1^*\|^2}{2} \leq c_4 \, \varepsilon^4; \quad \|x_1^*\| \leq \sqrt{\frac{2 \, c_4}{m}} \, \varepsilon^2 = c_3 \, \varepsilon^2.$$

Let us observe that $c_3$ is defined by the constants $L$, $m$ and $M$. Observing that the search will be continued within the hyperplane $L(x_0)$ of dimension $p$, we deduce from the inductive assumption that

$$\|x_n\| \leq \|x_{p+1}\| \leq \|x_1^*\| + \|x_{p+1} - x_1^*\| \leq c_3 \, \varepsilon^2 + c_p \, \|x_1 - x_1^*\|^2.$$

By assumption, the constant $c_p$ is defined only by the constants $L$, $M$, $m$ and $p$ provided $\|x_1 - x_1^*\|$ is sufficiently small. We shall estimate $\|x_1 - x_1^*\|$.

Since $f(x_0) > f(x_1) \geq f(x_1^*) \geq f(0) = 0$, we have

$$f(x_1) - f(x_1^*) < f(x_0) \leq \frac{M \, \varepsilon^2}{2} \, .$$

On the other hand,

$$\frac{m \, \|x_1 - x_1^*\|}{2} \leq f(x_1) - f(x_1^*).$$

The two preceding inequalities yield

$$\|x_1 - x_1^*\| \leq \frac{M}{m} \, \varepsilon^2.$$

Consequently,

$$\|x_n\| \leq \left(c_3 + c_p \frac{M}{m}\right) \varepsilon^2 = c \, \varepsilon^2,$$

where the constant $c$ depends only on $L$, $m$, $M$ and $n$. The proof is complete.  $\square$

We shall now present a version of the $r$-algorithm with no resetting. Although strict results on the rate of its convergence have not been obtained yet, numerical experiments indicate that the convergence is rapid in comparison with various variable metric methods and the conjugate gradient method [83]. The essential feature, however, is that some versions of $r$-algorithms can be used for minimizing nonsmooth functions, whereas the other rapidly convergent methods require at least the continuity of the gradient.

The $r$-algorithm with a constant coefficient of space dilation was implemented on the BESM-6 computer. We shall briefly describe the code, focusing our attention on the procedure for finding a minimum of $f$ in a direction $\eta$ and thus determining the stepsize $h_{k+1}$. Because directional minimization with high accuracy requires many function evaluations, in the implemented code line searches were organized in such a way that the accuracy of minimum seeking increased in the course of computation.

In all the numerical examples below the line search was performed according to the following procedure.

We choose numbers $\mu \geqq 1$ and $\gamma, 0 < \gamma < 1$, an integer $L$ and an initial trial stepsize $h$.

At iteration $k + 1$ we have a current point $x_{k+1}$, a direction $\eta$ in which we shall proceed from $x_k$ and a trial stepsize $h$.

From the point $x_k$ we make a step of length $h$ in the direction $\eta$, thus obtaining $z_1 = x_k + h\,\eta$. We evaluate $f(z_1)$. If $f(z_1) < f(x_k)$ then from $z_1$ we again advance in the direction $\eta$ and get $z_2 = z_1 + h\,\eta$. If $f(z_2) < f(z_1)$ we proceed further in this manner until an increase of the function value is observed. The last point thus obtained is taken as $x_{k+1}$. Let $l$ be the number of steps performed by the above procedure. If $l > L$ then we set the new trial stepsize equal to $\mu\,l\,h/L$. In case of the first step being unsuccessful the next trial stepsize is set equal to $\gamma\,h$. In all other cases the trial stepsize remains unchanged. In the examples below the parameters $\gamma, \mu$ and the trial stepsize were set equal to 0.1, 1.25 and 0.1, respectively.

The above line search procedure does not ensure that the function values decrease at each iteration, but numerical experiments show that increases occur rarely.

The program monitors the magnitude of the entries $b_{ij}$ of the matrix $B$ $(B = A^{-1}$, where $A$ is the matrix of space transformation). After every $p$ iterations the quantity $d = \max_{i,j} |b_{ij}|$ is calculated and the matrix $B$ is multiplied by a constant $\varrho$ whenever $d < \varepsilon$. The computations were performed with $p = 10$, $\varepsilon = 1$ and $\varrho = 10$.

Let us discuss the computational complexity of the algorithm. The basic operations in the code are the multiplication of a vector by a matrix, which requires $2n^2$ multiplications and additions ($n$ is the dimension of the space), and the multiplication of the matrix $B$ by the space dilation matrix $R_{1/\alpha}(\xi)$. Owing to the special form of $R_\beta(\xi)$, the calculation of $BR_\beta(\xi)$ requires only $5\,n^2$ multiplications and additions.

Most of the storage required is taken up by the matrix $B$, which needs $n^2$ cells. The memory of the BESM-6 computer allowed for solving problems of dimension up to 140.

In the description of the computational results we use the following notation: $f = f(t_1, \ldots, t_n)$ denotes the objective function, $x^*$ the exact minimum point of $f$, $x_0$ is the starting point, $\tilde{x}_N$ is the point obtained after $N$ iterations, and $\alpha$ denotes the coefficient of space dilation. In all examples the computations were stopped whenever one of the following tests was satisfied:

$$\|g_f(x_N)\| \leq 10^{-6}, \quad \|x_N - x_{N-1}\| \leq 10^{-7}, \quad \|B\, g_f(x_N)\| \leq 10^{-18}.$$

**Example 1** [66].

$$f(x) = 100\,(t_1^2 - t_2)^2 + (t_1 - 1)^2,$$
$$x^* = (1, 1), \quad f^* = 0, \quad x_0 = (-1.2, 1).$$

For $\alpha$ equal to 2 and 3 we have $N$ equal 63 and 39, respectively, and $\tilde{x}_N = (1.000000, 1.000000).$

**Example 2** [66].

$$f(x) = \sum_{i=1}^{10} (e^{-0.2i} + 2\,e^{-0.4i} - t_1\,e^{-0.2t_2 i} - t_3\,e^{0.4t_4 i})^2,$$
$$x^* = (1, 1, 2, 1), \quad f^* = 0, \quad x_0 = (0, 0, 0, 0);$$
$$\alpha = 2, \quad \tilde{x}_{83} = (0.999999, 1.000000, 2.000000, 1.000000),$$
$$\alpha = 3, \quad \tilde{x}_{90} = (0.999999, 0.999999, 2.000001, 1.000000).$$

**Example 3** [66].

$$f(x) = 10^{-3} \sum_{i=1}^{10} (10^3\,e^{-0.2i} + 2 \cdot 10^3\,e^{-0.4i} - t_1\,e^{-0.2t_2 i} - t_3\,e^{-0.4t_4 i})^2,$$
$$x^* = (1000, 1, 2000, 2), \quad f^* = 0, \quad x_0 = (500, 0, 2500, 3);$$
$$\alpha = 2, \quad \tilde{x}_{100} = (1000.000, 1.000000, 2000.000, 2.000000),$$
$$\alpha = 3, \quad \tilde{x}_{72} = (1000.001, 1.000000, 1999.999, 2.000000).$$

**Example 4** [95].

$$f(x) = 100\,(t_1^2 - t_2)^2 + (t_1 - 1)^2 + 90\,(t_3^2 - t_4)^2 + (t_3 - 1)^2$$
$$+ 10.1\,[(t_2 - 1)^2 + (t_4 - 1)^2] + 19.8\,(t_2 - 1)\,(t_4 - 1),$$
$$x^* = (1, 1, 1, 1), \quad f^* = 0, \quad x_0 = (-3, -1, -3, -1);$$
$$\alpha = 2, \quad \tilde{x}_{99} = (0.999999, 0.999999, 1.000000, 1.000000),$$
$$\alpha = 3, \quad \tilde{x}_{76} = (1.000000, 1.000000, 1.000000, 1.000000).$$

**Example 5** [95].

$$f(x) = (e^{t_1} - t_2)^4 + 100\,(t_2 - t_3)^6 + t_1^4\,(t_3 - t_4) + t_1^8 + (t_4 - 1)^2,$$

$$x^* = (0, 1, 1, 1), \quad f^* = 0, \quad x_0 = (1, 2, 2, 2);$$

$$\alpha = 2, \quad \tilde{x}_{36} = (0.0006, 1.0005, 1.0000, 1.0000),$$

$$\alpha = 3, \quad \tilde{x}_{32} = (-0.0004, 0.9996, 1.0000, 1.0000).$$

**Example 6** [95].

$$f(x) = (t_1 + 10\,t_2)^2 + 5\,(t_3 - t_4)^3 + (t_2 - 2\,t_3)^2 + 10\,(t_1 - t_4)^4;$$

$$x^* = (0, 0, 0, 0), \quad f^* = 0, \quad x_0 = (10, 10, 10, -10);$$

$$\alpha = 2, \quad \tilde{x}_{50} = (4 \cdot 10^{-7}, -2 \cdot 10^{-7}, 6 \cdot 10^{-7}, 7 \cdot 10^{-7}),$$

$$\alpha = 3, \quad \tilde{x}_{46} = (3.8 \cdot 10^{-5}, 0.4 \cdot 10^{-5}, -4.6 \cdot 10^{-5}, 4.6 \cdot 10^{-5}).$$

A detailed description of the numerical experiments can be found in [95].

The above results indicate that the $r$-algorithm is an efficient tool for solving manifold minimization problems with functions having distinct "gully" shapes. As far as the number of iterations required to reach the minimum with a given accuracy is concerned, our algorithm is close to the best versions of the Davidon-Fletcher-Powell method and the conjugate gradient method [66]. It is stable with respect to inexact line searches (a rather crude line search was used). Our numerical experience indicates that it is advisable to choose the space transformation coefficient $\alpha = 2 \div 3$ (the computations were performed for many other values of $\alpha$), since in most cases the decrease, if any, of the number of iterations, due to the further increase of $\alpha$, was insignificant and, at the same time, more computational effort at line searches was required (obviously, these recommendations apply to the implementation of the algorithm used).

A slightly modified version of the algorithm, in which exact directional minimizations were performed, was also tested numerically. A $20 \div 30$ per cent decrease in the number of iterations was observed, under identical stopping criteria.

## 3.7 Convergence of a Version of the $r$-Algorithm with Exact Directional Minimization

We shall now establish convergence of an "ideal" version of the $r$-algorithm which generates a minimizing sequence with monotonously decreasing function values [85].

Let $f$ be an almost differentiable function defined on the $n$-dimensional Euclidean space $E_n$ and let $G_f(x)$ denote the set of almost-gradients of $f$ at a point $x$ (the concepts of almost differentiable functions and almost-gradients were introduced in Sect. 1.4).

The following iterative procedure for minimizing $f$ will be called the $r_\mu(\alpha)$-*algorithm*, where $\alpha$ and $\mu$ are parameters satisfying $\alpha > 1$ and $0 \le \mu < 1$.

We initially have $x_0 \in E_n$, $\tilde{g}_0 = 0 \in E_n$ and a nonsingular $n \times n$ matrix $B_0$.

Suppose that after $k$ iterations $(k = 0, 1, \ldots)$ we have $x_k$, $\tilde{g}_k$ and $B_k$. On the $(k + 1)$-st iteration the following variables are calculated:

(1) $g_k^* = B_k^* g_f(x_k)$, where $g_f(x_k) \in G_f(x_k)$ is such that

$$(B_k^* g_f(x^k), \tilde{g}_k) \le \mu \, \| B_k^* g_f(x_k) \| \, \| \tilde{g}_k \| ,$$

i.e. $g_k^*$ is an almost-gradient of the function $\varphi_k(y) = f(B_k y)$ at the point $y_k = A_k x_k$, where $A_k = B_k^{-1}$;

(2) $r_k = g_k^* - \tilde{g}_k$, i.e. $r_k$ is the difference of two almost-gradients of the function $\varphi_k(y)$ calculated at the points $y_k = A_k x_k$ and $\tilde{y}_k = A_k x_{k-1}$;

(3) $\xi_{k+1} = \dfrac{r_k}{\| r_k \|}$;

(4) $B_{k+1} = B_k R_\beta(\xi_{k+1})$, i.e. $B_{k+1}$ is the inverse of the matrix $A_{k+1}$ which defines the transformation of the space after the $(k + 1)$-st iteration:

$$A_{k+1} = R_\alpha(\xi_{k+1}) A_k,$$

where $R_\alpha(\xi)$ is the operator of space dilation in direction $\xi$ satisfying $\| \xi \| = 1$, $\alpha$ is the coefficient of dilation of the space $E_n$, $\beta = 1/\alpha$ is the coefficient of "contraction" of the space of gradients;

(5) $\tilde{g}_{k+1} = R_\beta(\xi_{k+1}) g_k^* = B_{k+1}^* g_f(x^k)$, i.e. $g_{k+1}$ is an almost-gradient of the function $\varphi_{k+1}(y) = f(B_{k+1} y)$ at the point $\tilde{y}_{k+1} = A_{k+1} x_k$;

(6) $\qquad x_{k+1} = x_{k+1}(h_{k+1}) = x_k - h_{k+1} B_{k+1} \tilde{g}_{k+1},$ $\qquad\qquad$ (3.48)

where $h_{k+1}$ is chosen to satisfy
(a) the function $\psi(h) = f(x_{k+1}(h))$ is nonincreasing on the segment $[0; h_{k+1}]$, and
(b) there exists $g \in G_f(x_{k+1})$ such that

$$(B_{k+1}^* g, g_{k+1}) \le \mu \, \| B_{k+1}^* g \| \, \| \tilde{g}_{k+1} \| . \qquad\qquad (3.49)$$

Let us apply the operator $A_{k+1}$ to both sides of formula (3.48):

$$y_{k+1} = A_{k+1} x_{k+1} = A_{k+1} x_k - h_{k+1} \tilde{g}_{k+1} = \tilde{y}_{k+1} - h_{k+1} \tilde{g}_{k+1}.$$

Thus formula (3.48) in fact realizes one step of the gradient descent for the function $\varphi_{k+1}(y)$. When $\mu = 0$ we obtain an analogue of the iteration of the method of steepest descent.

The last step of the $(k + 1)$-st iteration is the following:

(7) the algorithm proceeds to the $(k + 2)$-nd iteration storing $x_{k+1}$, $\tilde{g}_{k+1}$ and $B_{k+1}$, or terminates if some stopping criterion is satisfied.

In contrast with other gradient-type algorithms for minimizing nondifferentiable functions (such as the subgradient method or the SDG method), the $r_\mu(\alpha)$-algorithm ensures a monotone descent $f(x_0) \geq f(x_1) \geq \ldots \geq f(x^k) \geq \ldots$, owing to the stepsize selection rules (3.48) and (3.49). At the same time, it differs substantially from the usual descent methods in that $h_{k+1} = 0$ does not imply a termination of descent. While executing a series of iterations with null steps, $x_k$ remains constant, but $B_k$ and $\tilde{g}_k$ vary. Thus an implicit search for a suitable direction of descent occurs.

In order to exclude from consideration some pathological cases, we shall assume that $f$ is a piecewise smooth function.

Let us define in detail the class $K$ of almost differentiable piecewise smooth functions. Let a partition of $E_n$ into $m$ closed sets $\bar{D}_1, \bar{D}_2, \ldots, \bar{D}_m$ be given such that their interiors $D_i$ are homeomorphic to an open sphere or an open halfspace and are mutually disjoint. We assume that there exist continuous and continuously differentiable functions $f_i$ defined on open sets $D_i^+ \supset \bar{D}_i, i = 1, 2, \ldots, m$.

We say that $f \in K$, i.e. $f$ is a continuous piecewise smooth function formed from functions $f_i$ on $\bar{D}_i, i = 1, \ldots, m$ if

(a) $f(x) = f_i(x)$ for all $x \in D_i$,
(b) $f(x) = f_i(x) = f_j(x)$ for all $x \in \bar{D}_i \cap \bar{D}_j$ and $i, j = 1, \ldots, m$.

One easily concludes that $f$ is almost differentiable. At each $x \in D_i$ the set $G_f(x)$ consists of a single element $g_{f_i}(x)$, which is the gradient of $f_i$ at $x$. If $x$ belongs to a common boundary of the domains of smoothness (pieces) with indices $i_1, \ldots, i_k$, where $k \geq 2$, then $G_f(x) = \{g_{f_{i_1}}(x), \ldots, g_{f_{i_k}}(x)\}$. For $f \in K$ an evaluation of an almost gradient consists of gradient evaluations of the functions from which $f$ is formed.

From now on we shall assume that

$$\lim_{x \to \infty} f(x) = +\infty. \tag{3.50}$$

Under this assumptions steps 1 and 6 of the $r_\mu(\alpha)$-algorithm can always be executed. In fact, if in step 6 one searches for a local minimum along the direction $-B_{k+1}\tilde{g}_{k+1}$ (in view of (3.50) such a minimum always exists), then property (a) is trivially satisfied; as for property (b), it suffices to show that there exists $g \in G_k(x_{k+1})$ satisfying $(B_{k+1}^* g, \tilde{g}_{k+1}) \leq 0$. But if for all $g \in G_f(x_{k+1})$ we had $0 < (B_{k+1}^* g, \tilde{g}_{k+1}) = (g, B_{k+1}\tilde{g}_{k+1})$, then $x_{k+1}$ could not be a local minimum in the direction $-B_{k+1}\tilde{g}_{k+1}$. Thus $x_{k+1}$ satisfies also property (b). The almost gradient, whose existence is stipulated in step 6, is used in step 1 of the next iteration.

**Remark.** The introduction of the constant $\mu$ instead of zero enables us to consider algorithms that use inexact line searches.

To study the $r_\mu(\alpha)$-algorithms we shall need some auxiliary results on the change of the width of convex sets under the operation of space dilation (contraction).

Let $W$ be a convex, closed and bounded body in $E_n$, and let $\eta$ be a unit vector, $\|\eta\| = 1$. Consider hyperplanes of the form $(\eta, x) - d = 0$. Denote

$$d_\eta^- = \max\{d \,|\, (\eta, x) - d \geq 0, \; x \in W\},$$
$$d_\eta^+ = \min\{d \,|\, (\eta, x) - d \leq 0, \; x \in W\}. \tag{3.51}$$

Clearly, $d_\eta^+ \geq d_\eta^-$. The width of the body $W$ in the direction $\eta$ is the number $d_\eta(W) = d_\eta^+ - d_\eta^-$. Geometrically, this is the distance between the two parallel hyperplanes that support the body $W$, have a common normal vector $\eta$ and are such that $W$ lies between them. The width $d(W)$ of the body $W$ is defined as $\min\limits_{\|\eta\|=1} d_\eta(W)$, and the diameter $D(W)$ of $W$ is $\max\limits_{\|\eta\|=1} d_\eta(w)$. It is known that

$$D(W) = \sup_{x, y \in W} \|x - y\|.$$

We shall now analyze the change of the width of a convex body under linear transformation of the space.

**Lemma 3.2.** *Let $B$ be a linear operator with a polar decomposition $B = SO$, where $O$ is an orthogonal operator and $S$ is a nonnegative definite symmetric operator with a minimum eigenvalue $\lambda(B)$. Then the width of the set $BW$ (the image of $W$ under the transformation $B$) satisfies the inequalities*

$$\lambda(B)\, d(W) \leq d(BW) \leq \lambda(B)\, D(W).$$

*Proof.* Let $x_1$ and $x_2$ be two boundary points of $W$ belonging to two supporting hyperplanes with the normal $\eta$ such that they are transformed by $B$ into the two hyperplanes which support $BW$ and whose distance is equal to the width of the body $BW$. Then

$$\|B(x_1 - x_2)\| = d(BW),$$

$$d(BW) \geq \min_{y \neq 0} \frac{\|By\|}{\|y\|} \, \|x_1 - x_2\| \geq \lambda(B)\, d_\eta(W) \geq \lambda(B)\, d(W).$$

Let $\bar\eta$, $\|\bar\eta\| = 1$, be a direction at which $\min\limits_{y \neq 0} \dfrac{\|By\|}{\|y\|}$ is attained, i.e. $\|B\bar\eta\| = \lambda(B)$. Then

$$d_{B\bar\eta/\|B\bar\eta\|}(BW)\, \lambda(B)\, d_{\bar\eta}(W) \leq \lambda(B)\, D(W).$$

This completes the proof. $\quad\square$

**Lemma 3.3.** *Let $z_1 \in W$, $z_2 \in W$, $\beta \leq 1$, $\gamma = \|z_1 - z_2\|/d(W) \geq 1$, and let $R_\beta(\xi)$ denote the operator of space dilation in the direction $\xi$ with the coefficient $\beta$. Then*

$$d\left(R_\beta\left(\frac{z_1 - z_2}{\|z_1 - z_2\|}\right) W\right) \geq \frac{d(W)}{\sqrt{1 + (1 - \beta^2)/(\beta^2 \gamma^2)}}.$$

*Proof.* Let $x_1$ and $x_2$ be two boundary points of $W$ belonging to two supporting hyperplanes $L_1$ and $L_2$ with the normal $\eta$, $\|\eta\| = 1$, which under the transformation $R_\beta((z_1 - z_2)/\|z_1 - z_2\|)$ are mapped into the two hyperplanes that support the body $R_\beta((z_1 - z_2)/\|z_1 - z_2\|)\,W$ and define its width. We have

$$d\left(R_\beta\left(\frac{z_1 - z_2}{\|z_1 - z_2\|}\right)W\right) = \left\| R_\beta\left(\frac{z_1 - z_2}{\|z_1 - z_2\|}\right)(x_1 - x_2)\right\|.$$

Consider a hyperplane passing through $x_1$ and $x_2$ and parallel to $z_1 - z_2$, and let $\bar{L}_1$ and $\bar{L}_2$ denote the straight lines formed by the intersection of that hyperplane with the hyperplanes $L_1$ and $L_2$, respectively. Let $z_1^0$ and $z_2^0$ be the points at which the straight line passing through $z_1$ and $z_2$ intersects $L_1$ and $L_2$, respectively, and let $\bar{z}_1$ and $\bar{z}_2$ be the images of $z_1$ and $z_2$ obtained by the parallel shift of the segment $z_1^0, z_2^0$ which transports $z_2^0$ to $x_2$. The image of $z_1^0$ under this shift will be denoted by $z$. Thus we have the geometric picture illustrated in the figure below.

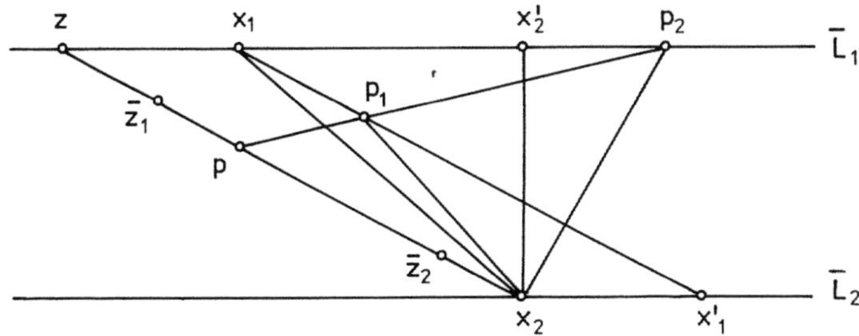


Let $x_2'$ be the projection of $x_2$ on $\bar{L}_1$, $x_1\,x_1' \parallel z\,x_2$, let $p$ denote the point of the segment $[z, x_2]$ such that $\|p - x_2\| = \beta\|z - x_2\|$, let $p_2$ be the intersection point of $\bar{L}_1$ with the straight line passing through $x_2$ and orthogonal to $[x_2, z]$, and let $p_1$ denote the intersection point of $[x_1, x_1']$ and $[p, p_2]$. By the definition of $R_\beta((z_1 - z_2)/\|z_1 - z_2\|)$, we have

$$\left\| R_\beta\left(\frac{z_1 - z_2}{\|z_1 - z_2\|}\right)(x_1 - x_2)\right\| = \|p_1 - x_2\|.$$

Let us find the minimum of $\|p_1 - x_2\|$ subject to the following constraints:

(1)   $\|z_1 - x_2\| \geqq \|z_1^0 - z_2^0\| = \|z_1 - z_2\| \geqq \gamma\,d(W),$

(2)   $\|x_2 - x_2'\| \geqq d_\eta(W) \geqq d(W).$

Clearly, the minimum is attained if $\|z - x_2\| = \gamma\,d(W)$, $\|x_2 - x_2'\| = d(W)$ and the segment $[x_2, p_1]$ is the height of the triangle $\triangle p\,p_2\,x_2$. By direct calculations we obtain

$$\min\|p_1 - x_2\| = \frac{d(W)}{\sqrt{1 + (1 - \beta^2)/(\beta^2\,\gamma^2)}},$$

which proves the lemma.   $\square$

For simplicity of exposition we shall assume in what follows that $\mu = 0$. The $r_0(\alpha)$-algorithm will be called the $r(\alpha)$-*algorithm*. For arbitrary $\mu$, $0 < \mu < 1$, the proofs proceed similarly to the case $\mu = 0$, with minor modifications.

Let $\{x_k\}_{k=0}^{\infty}$ denote a sequence of points generated by the $r(\alpha)$-algorithm used to minimize a function $f \in K$. Since $f(x_k) \leqq f(x_{k-1})$, if (3.50) holds then the limit

$$\lim_{k \to \infty} f(x^k) = f_\infty > -\infty$$

exists and the sequence $\{x_k\}_{k=0}^{\infty}$ is bounded.

Let $M_\infty$ denote the set of all accumulation points of the sequence $\{x_k\}_{k=0}^{\infty}$. Evidently, if $x \in M_\infty$ then $f(x) = f_\infty$. We are interested in the properties of the elements of $M_\infty$. Let us choose $\delta > 0$, $\varepsilon > 0$ and the ball $S_\delta(x)$ of radius $\delta$ centered at $x$.

Let $\bar{P}_{\delta, \varepsilon}(x)$ denote the convex closure of the set

$$P_{\delta, \varepsilon}(x) = (\bigcup_{y \in S_\delta(x)} G_f(y)) \cup (\bigcup_{z \in G_f(x)} S_\varepsilon(z)).$$

Let $W$ be a convex body. Let us introduce the function $e_\varrho(W) = \inf_{z \in W} |(\varrho, z)|$ for each $\varrho \in E_n$ satisfying $\|\varrho\| = 1$, and let $d_\varrho(W)$ denote the width of the set $W$ in the direction $\varrho$. Clearly, $d_\varrho(\bar{P}_{\delta, \varepsilon}(x)) \geqq \varepsilon$. Defining

$$K_\varrho(W) = \begin{cases} d_\varrho(W)/e_\varrho(W) & \text{if } e_\varrho(W) \neq 0, \\ +\infty & \text{if } e_\varrho(W) = 0, \end{cases}$$

we obtain

$$p(W) = \inf_{\|\varrho\| = 1} K_\varrho(W).$$

It is easy to show that $p(AW) = p(W)$ if $\det A \neq 0$.

**Theorem 3.11.** *Let $f \in K$, $\alpha > 1$, and let $\{x_k\}_{k=0}^{\infty}$ be a sequence constructed by the $r(\alpha)$-algorithm applied to $f$, satisfying*

$$\lim_{k \to +\infty} \|x_{k+1} - x_k\| = 0. \tag{3.52}$$

*Then for each fixed $v$, $\sqrt[n]{\beta} < v < 1$, $\varepsilon > 0$, $\delta > 0$ and a positive integer $r$ one can find $\bar{k} > r$ such that*

$$p(\bar{P}_{\delta, \varepsilon}(x_{\bar{k}})) \geqq \sqrt{\frac{v^2 \sqrt[n]{\alpha^2} - 1}{\alpha^2 - 1}}.$$

*Proof.* Suppose that for all $k > r$

$$p(\bar{P}_{\delta, \varepsilon}(x_k)) < \sqrt{\frac{v^2 \sqrt[n]{\alpha^2} - 1}{\alpha^2 - 1}}.$$

Let

$$D_{\max}^{(r)}(\delta,\,\varepsilon) = \sup_{k>r} D(\bar{P}_{\delta,\varepsilon}(x_k)).$$

It follows from (3.52) that for each positive integer $s$ there exists $N(s,\delta)$ such that $x_t \in S_\delta(x_k)$ for all $k > N(s,\delta)$ and $k \le t \le k+s$. Consequently, $g_f(x_k), g_f(x_{k+1}), \ldots, g_f(x_{k+s})$ belong to $\bar{P}_{\delta,\varepsilon}(x_k)$. Let us use Lemma 3.2 for estimating the width of the set $B_{k+1}^* \bar{P}_{\delta,\varepsilon}(x^k)$:

$$\varepsilon\,\lambda(B_{k+1}) \le d(B_{k+1}^* \bar{P}_{\delta,\varepsilon}(x_k)) \le \lambda(B_{k+1})\, D_{\max}^{(r)}(\delta,\,\varepsilon), \quad k > r.$$

By the rules of step 1 and step 2 of the $r$-algorithm, we have

$$r_{k+i} = B_{k+1}^* [g_f(x_{k+i}) - g_f(x_{k+i-1})],$$
$$(B_{k+i}^* g_f(x_{k+i}),\, B_{k+i}^* g_f(x_{k+i-1})) \le 0.$$

Therefore

$$\|r_{k+i}\| \ge \max \{\| B_{k+i}^* g_f(x_{k+i}) \|,\, \| B_{k+i}^* g_f(x_{k+i-1}) \|\},$$

$$\frac{d(B_{k+i}^* \bar{P}_{\delta,\varepsilon}(x^k))}{\|r_{k+i}\|} \le p(B_{k+i}^* \bar{P}_{\delta,\varepsilon}(x_k)), \quad i = 1,\ldots, s,$$

$$\frac{\|r_{k+i}\|}{d(B_{k+i}^* \bar{P}_{\delta,\varepsilon}(x^k))} \ge \frac{1}{p(B_{k+i}^* \bar{P}_{\delta,\varepsilon}(x_k))} \ge \sqrt{\frac{\alpha^2 - 1}{v^2 \sqrt[n]{\alpha^2} - 1}}\,. \tag{3.54}$$

Let us use Lemma 3.3 with

$$\beta = \frac{1}{\alpha}, \quad \gamma \ge \sqrt{\frac{\alpha^2 - 1}{v^2 \sqrt[n]{\alpha^2} - 1}}\,.$$

We obtain

$$d(B_{k+i+1}^* \bar{P}_{\delta,\varepsilon}(x_k)) = d(R_{1/\alpha}(\xi_{k+i+1})\, B_{k+i}^* \bar{P}_{\delta,\varepsilon}(x_k))$$

$$\ge d(B_{k+i}^* \bar{P}_{\delta,\varepsilon}(x_k)) \frac{1}{v\sqrt[n]{\alpha}} \quad (1 > v > \sqrt[n]{\beta}). \tag{3.55}$$

Let us choose $s$ so large that

$$\frac{1}{v^s} \ge \frac{D_{\max}^{(r)}(\delta,\,\varepsilon)}{\varepsilon}\,\bar{c} \tag{3.56}$$

where $\bar{c} > 1$. Then (3.54)−(3.56) yield

$$D_{\max}^{(r)}(\delta,\,\varepsilon)\,\lambda(B_{k+s+1}) \ge d(B_{k+s+1}^* \bar{P}_{\delta,\varepsilon}(x_k))$$

$$\ge d(B_{k+1}^* \bar{P}_{\delta,\varepsilon}(x_k)) \frac{\sqrt[n]{\beta^s}}{v^s}$$

$$\ge d(B_{k+1}^* \bar{P}_{\delta,\varepsilon}(x_k)) \frac{D_{\max}^{(r)}(\delta,\,\varepsilon)}{\varepsilon} \sqrt[n]{\beta^s}\, \bar{c}$$

$$\ge \lambda(B_{k+1})\, D_{\max}^{(r)}(\delta,\,\varepsilon) \sqrt[n]{\beta^s}\, \bar{c}\,.$$

Consequently,

$$\frac{\lambda(B_{k+s+1})}{\lambda(B_{k+1})} \geqq \bar{c}\sqrt[n]{\beta^s} \, .$$

The above argument remains valid for each $k > N(s, \delta)$. Therefore, for any $q$ and $\bar{k} > N(s, \delta)$ we have

$$\frac{\lambda(B_{\bar{k}+sq+1})}{\lambda(B_{\bar{k}+1})} \geqq \beta^{qs/n} \, \bar{c}^q, \quad q = 1, 2, \dots \; .$$

Then for sufficiently large $q$ one has $\lambda(B_{\bar{k}+qs+1}) > \beta^{(\bar{k}+qs+1)/n}$, which contradicts the fact that $\lambda(B_k) \leqq \beta^{k/n}$. The proof is complete. $\square$

**Theorem 3.12.** *Let the assumptions of Theorem* 3.11 *and condition* (3.50) *be satisfied. Then the set* $U = \{x : f(x) = f_\infty\}$ *contains a point* $x^*$ *such that the set of vectors* $G_f(x^*)$ *is linearly dependent.*

*Proof.* We shall prove the theorem by contradiction. Suppose that the set $G_f(x^*)$ consists of linearly independent vectors. Observe that the set $U$ is compact. Denote by $\Gamma(z)$ the Gramm determinant of the vectors belonging to $G_f(z)$. We shall show that

$$\inf_{z \in U} \Gamma(z) = e_0 > 0.$$

If this were not true, one could find a sequence $\{z_k\}_{k=1}^\infty$ satisfying

$$\lim_{k \to \infty} \Gamma(z_k) = 0$$

and such that all points $z_k$ would belong to the same piece or to the boundaries of the same fixed pieces $D_{i_1}, \dots, D_{i_r}$. But then, by the compactness of $U$, one could find a point $z^* \in U$ such that $\Gamma(z^*) = 0$, which would contradict the assumed linear independence of the vectors of $G_f(z^*)$.

Since $\inf\limits_{z \in U} \Gamma(z) = e_0 > 0$, there exists $e_1 > 0$ such that the Grammian of any subsystem of vectors in $G_f(z)$ is no less than $e_1$, for all $z \in U$.

We shall associate with each point $x \in U$ a ball neighborhood $S'_\varepsilon(x)$ such that: (1) it has common points with only those pieces $D_j$, $1 \leqq j \leqq m$, which are incident to $x$; (2) the Grammian of any subsystem of vectors in $G_f(z)$, $z \in S'_\varepsilon(x)$, is not less than $e_1/2$; (3) $\|g(z_1) - g(z_2)\| \leqq \varepsilon/2$, $\varepsilon > 0$, if the points $z_1, z_2 \in S'_\varepsilon(x)$ lie in the same piece. Let $S''(x)$ denote the open ball centered at $x$ and having the radius equal to a half of the radius of $S'_\varepsilon(x)$. Then

$$S = \bigcup_{x \in U} S''(x)$$

is an open set containing $U$. Since $U$ is a compact set, one may choose a finite subsystem of the neighborhoods $S''(z_1), \dots, S''(z_q)$, covering $U$. Let $\delta$ denote the smallest radius of those neighborhoods. Starting from a sufficiently large number $N$, each point $x_k$ of the minimizing sequence will lie inside some neighborhood $S''(z_{i_k})$, $1 \leqq i_k \leqq q$. Then the $\delta$-neighborhood of $x^k$ will be contained in $S'_\varepsilon(z_{i_k})$.

The distance from the origin to the hyperplane $\Pi(z)$ of minimum dimension which passes through all points of $G_f(x)$ is given by

$$\varrho[G_f(z)] = \left(\frac{\Gamma(G_f(z))}{\Gamma[g_2(z) - g_1(z), \ldots, g_{i_z}(z) - g_1(z)]}\right)^{1/2},$$

where $\{g_1(z), \ldots, g_{i_z}(z)\} = G_f(z)$. Since $\|g_2(z_{i_k}) - g_1(z_{i_k})\|, \ldots, \|g_{t_{z_{i_k}}}(z_{i_k}) - g_1(z_{i_k})\|$ are bounded and $\Gamma(G_f(z_{i_k})) \geqq e_1/2$, for all $k > N$ and sufficiently small $\varepsilon$ we have $e_\varrho(\bar{P}_{2\delta,\varepsilon}(z_{i_k})) = c > 0$, where $\varrho$ is the normal to $\Pi(z)$, $z = z_{i_k}$, and $d_\varrho(\bar{P}_{2\delta,\varepsilon}(z_{i_k})) \leqq 2\,\varepsilon$. Observing that $\bar{P}_{\delta,0}(x^k) \subset \bar{P}_{2\delta,0}(z_{i_k})$, for all $v > 0$ one can find $\varepsilon$ satisfying

$$p(\bar{P}_{\delta,\varepsilon}(x^k)) \leqq \frac{d_\varrho(\bar{P}_{2\delta,\varepsilon}(z_{i_k}))}{e_\varrho(P_{2\delta,\varepsilon}(z_{i_k}))} \leqq v \qquad \text{for all } k > N.$$

Since this contradicts Lemma 3.11, the proof is complete.  $\square$

From Theorem 3.11 one can obtain the following corollary.

**Theorem 3.13.** *Let $x^*$ be an isolated local minimum point of $f \in K$ and let $x_0$ be a starting point such that the set $S = \{x \in E_n : f(x^*) \leqq f(x) \leqq f(x_0)$ has a connected component containing $x^*$ and $x_0$. Suppose that this component does not contain, except for $x^*$, any other points $z$ with linearly dependent sets $G_f(z)$. Then, under the assumptions of Theorem 3.12, the sequence $\{x_k\}_{k=0}^\infty$ generated by the $r(\alpha)$-algorithm converges to $x^*$.*

We conclude with the following remarks. Theorem 3.13 provides sufficient conditions under which the $r(\alpha)$-algorithm converges to a local minimum. How stringent these conditions are? It is easy to observe that condition (3.52) is fulfilled, for instance, if the function $f$ has the following property: if $f$ decreases monotonically on the segment $[x_1, x_2]$ then for each $\varepsilon > 0$ one can find $\delta > 0$ such that $f(x) - f(y) > \varepsilon$ if $\|x - y\| > \delta$ and $x, y \in [x_1, x_2]$. This condition generalizes the property of uniform convexity. Unfortunately, it does not hold for linear functions. A typical situation in which the condition of linear independence of the family $G_f(z)$ is violated arises when $z$ belongs to more than $n$ pieces. As a rule, there is only a finite number of such points, and, as shown by our computational experience, if such points are not local minima, then they are not jamming points for the minimizing sequence.

Our numerical experiments in solving by the $r$-algorithm various problems of minimizing piecewise smooth functions [93, 6, 35, 104] have not so far detected any "pathological" example of divergence or of convergence to a point which is not a local minimum point. All the same it would be desirable to weaken the assumptions of Theorem 3.13. Also estimates of the speed of convergence in sufficiently general cases would be interesting.

## 3.8 Relations between SDG Algorithms and Algorithms of Successive Sections

Algorithms with space dilation, similar in structure to the SDG algorithms, have been recently shown to be closely related to methods of successive sections, and can be used for solving problems much more general than the problem of minimizing a convex function [87].

Assume we are given a vector field $g$ on $E_n$, which is not necessarily continuous; $g(x) \in E_n$ for all $x \in E_n$. We shall consider the problem of finding a point $x^*$ such that $(g(x), x - x^*) \geq 0$ for all $x \in E_n$. We assume that this problem has a solution; and that it is known a priori that $x^* \in S(x_0, R)$, where $S(x_0, R)$ denotes the closed ball with center $x_0$ and radius $R$.

Consider the following iterative algorithm for solving the problem in question for $n > 1$. In the description of the algorithm we assume that $g(x) \neq 0$ if $x \neq x^*$.

Before starting the computations, we have $x_0 \in E_n$, $B_0 = I_n$ (an identity matrix) and $h_0 = R/(n + 1)$. Suppose that at the $k$-th iteration we obtained $x_k \in E_n$, an $n \times n$ matrix $B_k$ and $h_k > 0$. At the $(k + 1)$-st iteration we calculate:

(1)   $g(x_k)$ (if $g(x_k) = 0$ then $x_k$ is a solution),

(2)   $\displaystyle \xi_k = \frac{B_k^* g(x_k)}{\| B_k^* g(x_k) \|}$,                                                        (3.57)

(3)   $x_{k+1} = x_k - h_k B_k \xi_k$,                                                              (3.58)

(4)   $B_{k+1} = B_k R_\beta(\xi_k)$,   $\displaystyle \beta = \sqrt{\frac{n - 1}{n + 1}}$,                          (3.59)

where $R_\beta(\xi_k)$ is the operator of space dilation in the direction $\xi_k$ with the coefficient $\beta$,

(5)   $h_{k+1} = r h_k$,   with   $\displaystyle r = \frac{n}{\sqrt{n^2 - 1}}$.                                  (3.60)

**Theorem 3.14.** *A sequence $\{x_k\}_{k=0}^\infty$ generated by the algorithm $(3.57) - (3.60)$ satisfies the inequalities*

$$\| A_k(x_k - x^*) \| \leq h_k(n + 1) \quad \text{for all } k = 0, 1, \ldots,$$                          (3.61)

*where $A_k = B_k^{-1}$.*

*Proof.* We shall prove the theorem by induction. For $k = 0$ inequality (3.61) reads $\| x_0 - x^* \| \leq R$ and holds by assumption. Suppose that (3.61) is satisfied for $k = \bar{k}$. We shall show that it also holds for $k = \bar{k} + 1$.

For brevity let $z_k = A_k(x_k - x^*)$ and $\alpha = 1/\beta$. Then

$$
\begin{aligned}
\| z_{\bar{k}+1} \|^2 &= \| R_\alpha(\xi_{\bar{k}})(z_{\bar{k}} - h_{\bar{k}}\xi_{\bar{k}}) \|^2 \\
&= \| z_{\bar{k}} - h_{\bar{k}}\xi_{\bar{k}} + (\alpha - 1)[(z_{\bar{k}} - h_{\bar{k}}\xi_{\bar{k}}, \xi_{\bar{k}})\,\xi_{\bar{k}}] \|^2 \\
&= \| z_{\bar{k}} + [-\alpha\, h_{\bar{k}} + (\alpha - 1)(z_{\bar{k}}, \xi_{\bar{k}})]\,\xi_{\bar{k}} \|^2 \\
&= \| z_{\bar{k}} \|^2 - 2 h_{\bar{k}}(z_{\bar{k}}, \xi_{\bar{k}})\,\alpha^2 + (\alpha^2 - 1)(z_{\bar{k}}, \xi_{\bar{k}})^2 + \alpha^2 h_{\bar{k}}^2 \\
&= \| z_{\bar{k}} \|^2 - 2 h_{\bar{k}}(z_{\bar{k}}, \xi_{\bar{k}})\,\frac{n+1}{n-1} + \frac{2}{n-1}(z_{\bar{k}}, \xi_{\bar{k}})^2 + \frac{n+1}{n-1}\,h_{\bar{k}}^2.
\end{aligned}
$$

Observe that

$$
\begin{aligned}
(z_{\bar{k}}, \xi_{\bar{k}}) &= \frac{1}{\| \tilde{g}(x_{\bar{k}}) \|}\,(A_{\bar{k}}(x_{\bar{k}} - x^*), B_k^* \, g(x_{\bar{k}})) \\
&= \frac{1}{\| \tilde{g}(x_{\bar{k}}) \|}\,(x_{\bar{k}} - x^*, g(x_{\bar{k}})) \geq 0,
\end{aligned}
$$

where $\tilde{g}(x_{\bar{k}}) = B_{\bar{k}}^* \, g(x_{\bar{k}})$, so $(z_{\bar{k}}, \xi_{\bar{k}}) \leq \| z_{\bar{k}} \| \leq h_{\bar{k}}(n+1)$. Therefore

$$
\frac{2}{n-1}(z_{\bar{k}}, \xi_{\bar{k}})^2 - 2 h_{\bar{k}}\frac{n+1}{n-1}(z_{\bar{k}}, \xi_{\bar{k}}) \leq \frac{2}{n-1}[(z_{\bar{k}}, \xi_{\bar{k}})^2 - (z_{\bar{k}}, \xi_{\bar{k}})^2] = 0.
$$

Hence

$$
\| z_{\bar{k}+1} \|^2 \leq \| z_{\bar{k}} \|^2 + \frac{n+1}{n-1}\,h_{\bar{k}}^2 \leq (n+1)^2 h_{\bar{k}}^2 + \frac{n+1}{n-1}\,h_{\bar{k}}^2
$$

$$
= (n+1)^2 \frac{n^2}{n^2-1}\,h_{\bar{k}}^2 \;=\; (n+1)^2 h_{\bar{k}+1}^2,
$$

and the desired conclusion follows. $\square$

The set of all points $x$ satisfying the inequality

$$
\| A_k(x - x^*) \| \leq (n+1)\,h_k = R\left(\frac{n}{\sqrt{n^2-1}}\right)^k
$$

is an ellipsoid $\Phi_k$ with volume $v(\Phi_k)$ equal to $v_0\,R^n\left(\dfrac{n}{\sqrt{n^2-1}}\right)^{nk}\Big/ \det A_k$

where $v_0$ is the volume of the unit $n$-dimensional ball. We obtain

$$
\begin{aligned}
\frac{v(\Phi_{k+1})}{v(\Phi_k)} &= \frac{\left(\dfrac{n}{\sqrt{n^2-1}}\right)^2 \det A_k}{\det A_{k+1}} = \frac{\left(\dfrac{n}{\sqrt{n^2-1}}\right)\det A_k}{\det R_\alpha(\xi_k)\,\det A_k} \\
&= \frac{1}{\alpha}\left(\frac{n}{\sqrt{n^2-1}}\right)^n = \frac{n-1}{n+1}\left(\frac{n}{\sqrt{n^2-1}}\right)^n = q_n < 1.
\end{aligned}
$$

The volume of the ellipsoid locating the solution $x^*$ according to inequality (3.61) decreases with the speed of a geometrical progression with the ratio $q_n$.

Let us consider possible applications of the above procedure to the solution of various mathematical programming problems.

## 1. The Problem of Minimizing a Convex Function on a Ball

Let $f$ be a convex function defined on $E_n$. We seek its minimum on the ball $S(x_0, R) = \{x: \|x - x_0\| \leq R\}$.

Let $x^*$ be a minimum point of $f$ on $S(x_0, R)$. Then for all $x \in S(x_0, R)$ we have

$$(g_f(x), x - x^*) \geq f(x) - f(x^*) \geq 0.$$

For any $x \notin S(x_0, R)$

$$(x - x_0, x - x^*) = \|x - x_0\|^2 - (x - x_0, x^* - x_0)$$
$$\geq \|x - x_0\| (\|x - x_0\| + \|x^* - x_0\|) \geq 0.$$

Thus we may construct a vector field $g$, satisfying $(g(x), x - x^*) \geq 0$ for all $x \in E_n$, as follows

$$g(x) = \begin{cases} g_f(x) & \text{if } x \in S(x_0, R), \\ \dfrac{x - x_0}{\|x - x_0\|} & \text{if } x \notin S(x_0, R). \end{cases} \tag{3.62}$$

Let us apply the algorithm $(3.57) - (3.60)$, with $g$ defined by $(3.62)$, to the problem in question. From Theorem 3.14 we obtain that the volume of the area locating the solution decreases with the speed of a geometrical progression with the ratio

$$q_n = \sqrt{\frac{n-1}{n+1}} \left( \frac{n}{\sqrt{n^2 - 1}} \right)^n.$$

Let us consider the case when the unconstrained minimum of $f$ is attained at an interior point $x^*$ of the ball $S(x_0, R)$. Since the algorithm $(3.57) - (3.60)$, when applied to the problem of minimizing $f$, in fact realizes the SDG algorithm with minor modifications, one may use Theorem 3.2 for estimating the speed of convergence of the function values. We obtain the following estimate

$$\varrho_k = \min_{i \in \overline{1,k}} \| B_i^* g(x_i) \| \leq \frac{d_k \sqrt{k(\alpha^2 - 1)}}{\sqrt{\alpha^{2k/n} - 1}},$$

$$d^k = \max_{i \in \overline{1,k}} \| g(x_i) \|. \tag{3.63}$$

Let the minimum in $(3.63)$ be attained at $i = i(k)$. Then

$$\mu_k = (x_{i(k)} - x^*, g(x_{i(k)})) = (A_{i(k)}(x_{i(k)} - x^*), B_{i(k)}^* g(x_{i(k)}))$$

$$\leq \| A_{i(k)}(x_{i(k)} - x^*) \| \frac{d_k \sqrt{k(\alpha^2 - 1)}}{\sqrt{\alpha^{2k/n} - 1}}$$

$$\leq h_{i(k)}(n + 1) \frac{d_k \sqrt{k(\alpha^2 - 1)}}{\sqrt{\alpha^{2k/n} - 1}}$$

$$\leq h_k(n + 1) d_k \frac{\sqrt{k(\alpha^2 - 1)}}{\sqrt{\alpha^{2k/n} - 1}} = \frac{R d_k \sqrt{k(\alpha^2 - 1)}}{\sqrt{1 - \beta^{2k/n}}} q_n^{k/n}.$$

We observe that if $x_{i(k)} \notin S(x_0, R)$ then $g(x_{i(k)}) = (x - x_0)/\|x - x_0\|$ and, since $x^*$ lies in the interior of $S(x_0, R)$, $(x_{i(k)} - x^*, (x_{i(k)} - x_0)/\|x_{i(k)} - x_0\|)$ $\geq \delta > 0$. Therefore, $\mu_k \to 0$ implies that $x_{i(k)} \in S(x_0, R)$ for all sufficiently large $k$. But if $x_{i(k)} \in S(x_0, R)$ then $g(x_{i(k)}) = g_f(x_{i(k)})$ and

$$f(x_{i(k)}) - f(x^*) \leq (g_f(x_{i(k)}), x_{i(k)} - x^*)$$

$$\leq \frac{R \, d_k \sqrt{k(\alpha^2 - 1)}}{\sqrt{1 - \beta^{2k/n}}} \, q_n^{k/n}.$$

Consequently, if $f$ attains its minimum at an interior point $x^*$ of the ball $S(x_0, R)$, then in the algorithm $(3.57) - (3.60)$ the deviation of the "record" function values from $f(x^*)$ satisfies

$$r_k = \min_{i \in \overline{1,k}} [f(x_i) - f(x^*)] \leq \frac{R \, d_k \sqrt{k(\alpha^2 - 1)}}{\sqrt{1 - \beta^{2k/n}}} \, q_n^{k/n}.$$

If $x_k \in S(x_0, R)$ for $k = 0, 1, \ldots$, then one can obtain from Theorem 3.1 the following more accurate estimate

$$f(x_{k_i}) - f(x^*) \leq c \, q_n^{k/n}$$

for some subsequence $\{k_i\}_{i=1}^{\infty}$. For large $n$, $q_n$ is well approximated by the following asymptotic formula

$$q_n \approx 1 - \frac{1}{2n}.$$

It follows that although $r_k$ converges essentially at the speed of a geometrical progression (if one neglects the slowly increasing multiplier $\sqrt{k}$), the coefficient of this progression is close to unity for large $n$, i.e. the convergence may be slow in practice.

An advantage of the algorithm $(3.57) - (3.60)$ is that the guaranteed speed of convergence depends only on the dimension of the space and does not require the knowledge of specific properties of $f$.

## 2. The General Convex Programming Problem

The problem is to

$$\text{minimize} f_0(x),$$

$$\text{subject to} f_i(x) \leq 0, \quad i = 1, \ldots, m, \quad x \in E_n, \quad n > 1, \tag{3.64}$$

where $f_v$ are convex functions defined on $E_n$ with subgradients $g_v(x)$, $v = 0, 1, \ldots, m$. We assume that it is known a priori that an optimal point $x^*$ exists in a ball $S(x_0, R)$ (formally, one may append the constraint $\|x - x_0\| \leq R$ to (3.64)).

Consider the vector field $g$

$$g(x) = \begin{cases} g_0(x) & \text{if } \max_{i \in \overline{1,m}} f_i(x) \leq 0, \\ g_{i*}(x) & \text{if } \max_{i \in \overline{1,m}} f_i(x) = f_{i*}(x) > 0. \end{cases} \tag{3.65}$$

We shall show that $(g(x), x - x^*) \geq 0$ for all $x \in E_n$. If $\max_{i \in \overline{1,m}} f_i(x) \leq 0$, then $g(x) = g_0(x)$ and

$$(g(x), x - x^*) = (g_0(x), x - x^*) \geq f_0(x) - f_0(x^*) \geq 0.$$

If $\max_{i \in \overline{1,m}} f_i(x) > 0$, then $g(x) = g_{i*}(x)$ with $f_{i*}(x) > 0$ and $f_{i*}(x^*) \leq 0$, so $(g(x), x - x^*) = (g_{i*}(x), x - x^*) \geq f_{i*}(x) - f_{i*}(x^*) \geq 0$. Therefore, $(g(x), x - x^*) \geq 0$ for all $x \in E_n$. In view of this inequality, we can use the algorithm $(3.57) - (3.60)$ for locating $x^*$, with $g(x)$ calculated by (3.65).

By Theorem 3.14, after $k$ iterations of the algorithm $(3.57) - (3.60)$ $x^*$ will be located in an ellipsoid $\Phi_k$ centered at $x_k$ with volume $v(\Phi_k) = v_0 q_n^k$ ($v_0$ denotes the volume of the ball $S(x_0, R)$).

Let us observe that the result remains valid if in (3.65) instead of $g_{i*}(x)$ one takes $g_{\bar{i}}(x)$ for an arbitrary index $\bar{i}$ satisfying $f_{\bar{i}}(x) > 0$.

## 3. The Saddle Point Problem

Let $f$ be a convex-concave function of two vector variables $x \in E_n$ and $y \in E_m$, $\{x, y\} = z \in E_n \times E_m = E_{n+m}$, let $z^*$ be a saddle point of $f$, let $z_0$ be an initial approximation to a solution and suppose that one knows *a priori* that $\|z_0 - z^*\| \leq R$.

Let us consider the pseudo-gradient set

$$G(z) = G_f^x(x, y) \times (- G_f^y(x, y)),$$

where $G_f^x(x, y)$ is the set of partial subgradients of the function $f(x, y)$, considered as a function of $x$ for fixed $y$, and $- G_f^y(x, y)$ is the set of subgradients of the function $- f(x, y)$ of $y$ for fixed $x$.

Define the vector field $g$ as follows

$$g(z) = \{g_f^x(z), - g_f^y(z)\}, \quad g_f^x(z) \in G_f^x(z), \quad g_f^y(z) \in G_f^y(z).$$

We shall show that $(g(z), z - z^*) \geq 0$.

By the definition of a saddle point, we have $f(x, y^*) \geq f(x^*, y^*) \geq f(x^*, y)$, so

$$\begin{aligned} 0 \leq f(x, y^*) - f(x^*, y) &= f(x, y^*) - f(x, y) + f(x, y) - f(x^*, y) \\ &\leq (g_f^x(z), x - x^*) - (g_f^y(z), y - y^*) \\ &= (g(z), z - z^*). \end{aligned}$$

Consequently, for locating $x^*$ one can use the algorithm $(3.57) - (3.60)$.

Other subgradient methods for saddle point seeking may be found in [39, 57].

One can derive algorithms of the form $(3.57)-(3.60)$ from the scheme of successive sections. A. Yu. Levin proposed in 1965 the *method of centered sections* for minimizing convex functions and solving convex programming problems. At each iteration of the method, knowing a closed convex polyhedron locating an optimal point, one finds the center of gravity of this polyhedron. By calculating a subgradient at this center one can locate the optimal point in the halfspace defined by the corresponding supporting hyperplane to the level set, and thus one can cut off the part of the polyhedron that lies in the other halfspace. It can be shown [53] that at each iteration at least an $(n/(n+1))^n$-th part of the volume of the polyhedron is cut off, which leads to a decrease in the location area with a speed not smaller than the speed of decrease of a geometrical progression with the ratio

$$Q_n = 1 - \left(\frac{n}{n+1}\right)^n \quad \left(Q_n \approx 1 - \frac{1}{e} \text{ for large } n\right).$$

However, for $n > 3$ this algorithm is impractical, since calculating centers of gravity of multidimensional polyhedra requires much effort. B. D. Yudin and A. S. Nemirovskii [103] proposed a modified method of centered sections for solving convex programming problems. Let an optimum point $x^*$ be located in a ball $S(x_0, R)$. Define the hyperplane $(g(x_0), x - x_0) = 0$. The location area of $x^*$ reduces to a half-ball. Let us construct an ellipsoid of minimum volume containing this half-ball. The center of the ellipsoid lies on the ray $x_0 - h\,g(x_0), h > 0$, and the ellipsoid is flattened in the direction $g(x_0)$. To complete the recursion, one has to transform the ellipsoid into a sphere, which can be done by space dilation in the direction $g(x_0)$ with a certain coefficient. Thus centered sections lead in a natural way to a gradient-type algorithm with space dilation in the direction of the gradient.

It is worth noting that one can use the cutting plane idea for constructing a family of algorithms of varying complexity. For instance, if the location area of $x^*$ is immersed in an ellipsoid after two cut-offs, then, in general, we need space dilation in two directions. One may expect that this will significantly increase the speed of convergence in comparison to that of the algorithm $(3.57)-(3.60)$.

## 3.9 Computational Modifications of Subgradient Methods with Space Dilation

In [96] V. A. Skokov proposed modified formulae for implementing subgradient methods with space dilation, which significantly reduce the number of arithmetic operations involved in space transformations. The essence of these modifications lies in updating the symmetric matrix $H_k = B_k B_k^*$ instead of $B_k$.

In the SDG algorithms as well as in the $r$-algorithms the step from $B_k$ to $B_{k+1}$ and from $x_k$ to $x_{k+1}$ proceeds according to formulae of the form

$$B_{k+1} = B_k R_\beta(\xi_{k+1}), \quad x_{k+1} = x_k - h_{k+1} B_{k+1} \bar{g}_{k+1},$$

where $\xi_{k+1}$ is of the form $B_k^* p_k / \| B_k^* p_k \|$, $p_k \in E_n$, and

$$\bar{g}_k = \frac{B_k^* g_f(x_k)}{\| B_k^* g_f(x_k) \|}.$$

Using formula (3.5), we obtain

$$H_{k+1} = B_{k+1} B_{k+1}^* = B_k R_\beta(\xi_{k+1}) R_\beta(\xi_{k+1}) B_k^* = B_k R_\beta^2(\xi_k) B_k^*$$

$$= B_k R_{\beta^2}(\xi_{k+1}) B_k^* = B_k (I + (\beta^2 - 1) \xi_{k+1} \xi_{k+1}^T) B_k^*$$

$$= H_k + (\beta^2 - 1) B_k \xi_{k+1} \xi_{k+1}^T B_k^*$$

$$= H_k + (\beta^2 - 1) \frac{B_k B_k^* p_k p_k^T B_k B_k^*}{(B_k^* p_k, B_k^* p_k)}$$

$$= H_k + (\beta^2 - 1) \frac{H_k p_k p_k^T H_k}{(H_k p_k, p_k)},$$

hence

$$H_{k+1} = H_k + \frac{H_k p_k p_k^T H_k}{(H_k p_k, p_k)} (\beta^2 - 1).$$

Then

$$x_{k+1} = x_k - h_{k+1} B_{k+1} \frac{B_{k+1}^* g_f(x_k)}{\| B_{k+1}^* g_f(x_k) \|}$$

$$= x_k - h_{k+1} \frac{H_{k+1} g_f(x_k)}{(H_{k+1} g_f(x_k), g_f(x_k))^{1/2}},$$

$$x_{k+1} = x_k - h_{k+1} \frac{H_{k+1} g_f(x_k)}{(H_{k+1} g_f(x_k), g_f(x_k))^{1/2}}.$$

Skokov's modifications enable one to use for storage, instead of $n^2$ memory cells required by the matrix $B_k$, only $n(n + 1)/2$ cells necessary for the symmetric matrix $H_k$. Each updating of the matrix $H_k$ requires approximately $2n^2$ multiplications. However, additional complications may arise from a possible loss of positive definiteness of the perturbed matrix $\tilde{H}_k$ resulting from the accumulation of numerical errors after a large number of iterations. With the original computational scheme the probability that the matrix $\tilde{H}_k \approx B_k B_k^*$ is not positive definite is much smaller. To increase the numerical stability of Skokov's updates certain regularizing modifications of $H_k$ are necessary; however, this question is still open.

# 4. Applications of Methods for Nonsmooth Optimization to the Solution of Mathematical Programming Problems

## 4.1 Application of Subgradient Algorithms in Decomposition Methods

Decomposition methods are used for solving large-scale linear and convex programming problems in order to save time by reducing the number of references to the external memory of a computer. Such methods convert the solution of the original problem into the solution of a series of problems of lower dimension (blocks). They are particularly efficient if the structure of each block permits the use of special, fast solution methods.

In mathematical programming two decomposition schemes, each dual to the other, are mainly considered: *decomposition with respect to variables* and *decomposition with respect to constraints*. More complex decomposition schemes are usually superpositions of the two basic ones.

Recently much attention has been devoted to the so-called parametric decomposition, in which one augments the original problem by additional variables (parameters), so that for fixed values of the parameters the structure of the problem simplifies or becomes a block structure.

The basic task of the solution process is to find optimal values of the parameters. Parametric decomposition may be regarded as a kind of decomposition with respect to variables. A large number of works are devoted to studying various decomposition schemes, mainly for linear programming problems [16, 71, 1].

We shall consider the use of the subgradient method in iterative algorithms for solving linear and convex programming problems with the aid of some decomposition schemes.

### Decomposition with Respect to Variables

Consider a convex programming problem having variables grouped in two subsets. The problem is to find

$$\min_{x,y} f_0(x, y), \tag{4.1}$$

subject to the constraints

$$f_i(x, y) \leq 0, \quad i = 1, \ldots, n, \tag{4.2}$$

where $x$ and $y$ are vector variables:

$$x = (x^{(1)}, \ldots, x^{(l)}), \quad y = (y^{(1)}, \ldots, y^{(m)}),$$

and $f_0$ and $f_i$, $i = 1, \ldots, n$, are convex functions. Let us fix $x = \bar{x}$ and consider the following problem: find

$$\min_y f_0(\bar{x}, y) \tag{4.3}$$

subject to the constraints

$$f_i(\bar{x}, y) \leq 0, \quad i = 1, \ldots, n. \tag{4.4}$$

It is easy to observe that problem $(4.3)-(4.4)$ is a convex programming problem. For those values of $\bar{x}$ for which a solution to problem $(4.3)-(4.4)$ exists, we shall define the function $\Phi$ as follows:

$$\Phi(\bar{x}) = \min_{y \in D(\bar{x})} f_0(\bar{x}, y), \tag{4.5}$$

where $D(\bar{x})$ is the set of all $y$ satisfying (4.4).

**Theorem 4.1.** *If $f_\alpha$, $\alpha = 0, 1, \ldots, n$, are convex functions, then the function $\Phi$ defined by (4.5) is convex on some convex subset $W$ of $E_n$. If for some $\bar{x} \in W$ the Slater constraint qualification is fulfilled for (4.4), then a subgradient of $\Phi$ at a point $x = \bar{x}$ can be calculated by the formula*

$$g_\Phi(\bar{x}) = g_{L_U}^{\bar{x}}(\bar{x}, y(\bar{x})), \tag{4.6}$$

*where*

$$L_U(x, y) = f_0(x, y) + \sum_{i=1}^n U_i f_i(x, y),$$

*$y(\bar{x})$ is one of the optimal values of $y$ in problem $(3.4)-(4.4)$, $U = \{U_i\}$ are Lagrange multipliers of $(4.3)-(4.4)$ obtained from the Kuhn-Tucker theorem, and $g_{L_U}^{\bar{x}}(\bar{x}, y(\bar{x}))$ is the projection of a subgradient of the function $L_U(z) = L_U(x, y)$ on the subspace $E_l^x$ which has a null projection on the subspace $E_m^y$ (the subgradient is taken at the point $z = (\bar{x}, y(\bar{x})))$.*

*Proof.* Consider two points $\bar{x}_1$ and $\bar{x}_2$ and their convex combination

$$\bar{x}_3 = \lambda_1 \bar{x}_1 + \lambda_2 \bar{x}_2 \quad (\lambda_1 + \lambda_2 = 1; \lambda_1, \lambda_2 \geq 0).$$

Let $\bar{y}_1 = y(\bar{x}_1)$ and $\bar{y}_2 = y(\bar{x}_2)$ be two solutions of $(4.3)-(4.4)$ for $\bar{x} = \bar{x}_1$ and $\bar{x} = \bar{x}_2$, respectively, and let $\bar{y}_3 = \lambda_1 \bar{y}_1 + \lambda_2 \bar{y}_2$. Then, by the convexity of $f_\alpha$ $(\alpha = 0, 1, \ldots, n)$, we have

$$f_0(\bar{x}_3, \bar{y}_3) \leq \lambda_1 f_0(\bar{x}_1, \bar{y}_1) + \lambda_2 f_0(\bar{x}_2, \bar{y}_2),$$
$$f_i(\bar{x}_3, \bar{y}_3) \leq \lambda_1 f_i(\bar{x}_1, \bar{y}_1) + \lambda_2 f_i(\bar{x}_2, \bar{y}_2) \leq 0.$$

Therefore $\bar{y}_3$ is feasible for problem $(4.3)-(4.4)$ for $\bar{x} = \bar{x}_3$. Consequently,

$$\Phi(\bar{x}_3) = \Phi(\lambda_1 \bar{x}_1 + \lambda_2 \bar{x}_2) \leq \lambda_1 \Phi(\bar{x}_1) + \lambda_2 \Phi(\bar{x}_2),$$

which proves the convexity of $\Phi$.

We now pass to the proof of formula (4.6) for the subgradient of $\Phi$. By the Kuhn-Tucker theorem, if $D(\bar{x})$ satisfies the Slater condition then there exist multipliers $U(\bar{x}) = \{U_i(\bar{x})\}$, $i = 1, \ldots, n$, such that the following condition is fulfilled

$$\Phi(\bar{x}) = \min_y [f_0(\bar{x}, y) + \sum_{i=1}^n U_i(\bar{x}) f_i(\bar{x}, y)]$$

$$= \max_{U \geq 0} \min_y [f_0(\bar{x}, y) + \sum_{i=1}^n U_i f_i(\bar{x}, y)].$$

If the Slater condition holds for $D(\bar{x})$, then it also holds for $D(x)$ if $x$ belongs to some neighborhood $S(\bar{x})$ of $\bar{x}$ (by the continuity of $f_i$). For $x \in S(\bar{x})$ we have

$$\Phi(x) = f_0(x, y(x)) + \sum_{i=1}^n U_i f_i(x, y(x))$$

$$\geq f_0(x, y(x)) + \sum_{i=1}^n \bar{U}_i f_i(x, y(x))$$

$$\geq L_{\bar{U}}(x, y),$$

where $\bar{U} = \{\bar{U}_i\}$ are the Lagrange multipliers of $(4.3)-(4.4)$ at $x = \bar{x}$. Let us choose a subgradient of $L_U$ at $(x, y) = (\bar{x}, y(\bar{x}))$ such that its projection $g_L^y(\bar{x}, y(\bar{x}))$ on the subspace $E_m^y$ vanishes (this is always possible, since

$$L_U(\bar{x}, y(\bar{x})) = \min L(\bar{x}, y),$$

which means that the subdifferential of $L_U$ at $(\bar{x}, y(\bar{x}))$ intersects the hyperplane $y = 0$). We obtain

$$\Phi(x) - \Phi(\bar{x}) \geq L_{\bar{U}}(x, y(x)) - L_{\bar{U}}(\bar{x}, y(\bar{x}))$$

$$\geq (x - \bar{x}, g_{L_U}^x(\bar{x}, y(\bar{x}))) + (y(x) - y(\bar{x}), g_{L_U}^y(\bar{x}, y(\bar{x})))$$

$$= (x - \bar{x}, g_{L_U}^x(\bar{x}, y(\bar{x}))),$$

which yields (4.6) by the definition of a subgradient. $\quad\square$

**Corollary.** *If in the preceding lemma one additionally assumes $f_\alpha(x, y)$, $\alpha = 0, 1, \ldots, n$, to be continuously differentiable with respect to $y$, then formula (4.6) can be made more precise:*

$$g(\bar{x}) = g_{f_0}^x(\bar{x}, y(\bar{x})) + \sum_{i=1}^n U_i(\bar{x}) g_{f_i}^x(\bar{x}, y(\bar{x})), \tag{4.7}$$

*where $g_{f_\alpha}^x(\bar{x}, y(\bar{x}))$ are projections of arbitrary subgradients of $f$ at $\bar{z} = (\bar{x}, y(\bar{x}))$ on $E_l^x$, $\alpha = 0, 1, \ldots, n$.*

*Proof.* This result follows from the fact that in the case of continuous differentiability of $f$ with respect to $y$, the projection $g^y_{L_v}(\bar{x}, y(\bar{x}))$ is uniquely determined and null. It remains to apply the formula representing a subgradient of the function $L_U$, which is a nonnegative combination of $f_\alpha$, in terms of subgradients of the forming functions (see Sect. 1.3).   $\square$

Formula (4.7) enables one to easily construct algorithms for solving convex programming problems (in particular, linear programs) by applying decomposition with respect to variables. To this end we have to assume that it is possible to solve problem (4.3)−(4.4) in a finite number of steps, obtain Lagrange multipliers $U = \{U_i\}$, and compute "partial" subgradients $g^x_{f_\alpha}$, $\alpha = 0, \ldots, n$. These assumptions can be fulfilled, for instance, if (4.3)− (4.4) is a linear or quadratic programming problem. Additional complications may arise from the fact that problem (4.3)−(4.4) may have no solutions for some $\bar{x}$. A standard way out of this difficulty is to use the method of penalty function, in which problem (4.1)−(4.2) is replaced by the following one: find

$$\min_{x,y,v} f_0(x, y) + M \sum_{i=1}^{n} v_i \tag{4.8}$$

subject to the constraints

$$f_i(x, y) - v_i \leqq 0, \quad v_i \geqq 0, \tag{4.9}$$

where $M$ is a sufficiently large positive number. Then problem (4.3)−(4.4) is replaced by the problem: find

$$\min_{y,v} \left[ f_0(\bar{x}, y) + M \sum_{i=1}^{n} v_i \right], \tag{4.10}$$

subject to the constraints

$$f_i(\bar{x}, y) - v_i \leqq 0, \quad v_i \geqq 0. \tag{4.11}$$

Clearly, for any $\bar{x}$ problem (4.10)−(4.11) has a nonempty feasible set which satisfies the Slater condition.

Thus we may suppose that problem (4.1)−(4.2) is reduced to a form in which the function $\Phi(\bar{x})$ is defined for all $\bar{x}$ and one can compute a subgradient $g_\Phi(\bar{x})$ at any $\bar{x}$, i.e. one may use the subgradient algorithm for minimizing $\Phi$. Consequently, we obtain the following algorithm for solving problem (4.1)−(4.2), whose $k$-th iteration consists of the following steps:

(a) for the previously calculated $\bar{x} = \bar{x}_k$ solve problem (4.3)−(4.4) and find the Lagrange multipliers $U(\bar{x}_k) = \{U_i(\bar{x}_k)\}$,
(b) calculate the vector $g_\Phi(\bar{x}_k)$ by (4.6), or, in favourable cases, by (4.7),
(c) compute $\bar{x}_{k+1} = \bar{x}_k - h_{k+1} g_\Phi(\bar{x}_k)$.

The stepsizes $h_k$ are selected according to the analysis and recommendations of Chap. 2.

For the special case of $(4.1)-(4.2)$ being a linear programming problem, we shall describe the above algorithm in a more detailed way. Suppose one wants to find

$$\max\,[(c,\,x)+(d,\,y)] \tag{4.12}$$

subject to the constraints

$$A\,x+B\,y\leqq e, \tag{4.13}$$

$$x\geqq 0,\quad y\geqq 0, \tag{4.14}$$

where $x$ and $y$ are vector variables of dimension $l$ and $m$, respectively, $A$ is an $n\times l$-matrix, $B$ is an $n\times m$ matrix; $c,\,d$ and $e$ are fixed vectors of dimension $l,\,m$ and $n$, respectively. Let us fix $x=\bar{x}$ and consider the problem: find

$$\max_{y}\,[(c,\,\bar{x})+(d,\,y)], \tag{4.15}$$

subject to the constraints

$$B\,y\leqq e-A\,\bar{x}, \tag{4.16}$$

$$y\geqq 0 \tag{4.17}$$

and its dual: find

$$\min_{v}(e-A\,\bar{x},\,v), \tag{4.18}$$

subject to the constraints

$$B^{*}\,v\geqq d, \tag{4.19}$$

$$v\geqq 0, \tag{4.20}$$

where $v$ is a variable vector of dimension $n$, $B^{*}$ is the transpose of the matrix $B$.

Let $\bar{v}(\bar{x})$ denote a solution of $(4.18)-(4.20)$ and $\alpha_{j}$ denote the $j$-th row of the matrix $A$ $(j=1,\ldots,l)$. Define the $j$-th component of the $l$-vector $g(\bar{x})$ as follows:

$$g_{j}(\bar{x})=(-\,\alpha_{j},\,\bar{v}(\bar{x}))+c_{j},\quad g(\bar{x})=(g_{1}(\bar{x}),\ldots,g_{l}(\bar{x})), \tag{4.21}$$

where $c_{j}$ is the $j$-th component of the vector $c$. Then the iterative algorithm for solving problem $(4.12)-(4.14)$ according to the above decomposition scheme can be described as follows. We choose an initial point $\bar{x}_{0}$, and after $k$ iterations obtain a point $\bar{x}_{k}$.

On the $(k+1)$-st iteration we proceed as follows:
(1) we solve problem $(4.18)-(4.20)$ at $\bar{x}=\bar{x}_{k}$ and find the vector $v(\bar{x}_{k})$,
(2) we compute the subgradient $g(\bar{x}_{k})$ by formula (4.21) with $\bar{x}=\bar{x}_{k}$,
(3) we calculate $\bar{x}_{k+1}$ by the formula

$$\bar{x}_{k+1}=\max\,\{0,\,\bar{x}_{k}+h_{k+1}\,g(\bar{x}_{k})\}, \tag{4.22}$$

where $h_{k+1}$ is the stepsize and "max" denotes coordinatewise maximization.

The decomposition method described above can be used for solving various optimization problems and for analysing large transportation networks.

**Example 1.** Determination of optimal potentials in the transportation problem.

Let us consider the problem, which is a dual of the transportation problem: find

$$\max \left[ \sum_{j=1}^{m} b_j y_j - \sum_{i=1}^{l} a_i x_i \right], \tag{4.23}$$

subject to the constraints

$$y_j - x_i \leqq c_{ij}; \quad i = 1, \ldots, l; \quad j = 1, \ldots, m. \tag{4.24}$$

We suppose that $\sum_{i=1}^{l} a_i = \sum_{j=1}^{m} b_j$.

We have two groups of variables: $x = \{x_i\}_{i=1}^{l}$ and $y = \{y_i\}_{i=1}^{m}$. Using the above decomposition scheme, we proceed as follows. Let us fix $x = \bar{x} = \{\bar{x}_i\}$ and consider the problem of finding

$$\max_{y} \left[ \sum_{j=1}^{m} b_j y_j - \sum_{i=1}^{l} a_i x_i \right], \tag{4.25}$$

subject to the constraints

$$y_j \leqq c_{ij} - \bar{x}_i, \tag{4.26}$$

and its dual problem: find

$$\min_{v} \sum_{i=1}^{l} \sum_{j=1}^{m} (c_{ij} + \bar{x}_i)\, v_{ij}, \tag{4.27}$$

subject to

$$\sum_{i=1}^{l} v_{ij} = b, \tag{4.28}$$

$$v_{ij} \geqq 0. \tag{4.29}$$

Problems $(4.25)-(4.26)$ and $(4.27)-(4.29)$ have trivial solutions for $b_j \geqq 0$:

$$v_{ij}(\bar{x}) = \begin{cases} y_j(\bar{x}) = \min_{i} (\bar{x}_i + c_{ij}) \\ b_j \text{ for } i = i^*(j) \text{ such that } \bar{x}_{i^*(j)} + c_{ij} = y_i(\bar{x}), \\ 0 \text{ for other combinations of indices.} \end{cases}$$

Therefore

$$g_i(\bar{x}) = \sum_{j \in M_i} b_j - a_i,$$

where $M_i$ is the set of indices $j$ satisfying $i^*(j) = i$.

In this way a formal application of the decomposition scheme described above leads to a well-known algorithm for solving the transportation problem by the method of subgradient descent in the space of potentials [75].

**Example 2.** The multicommodity transportation problem [92]. Consider a transportation network $s$ having vertices $i$ and arcs $d_j$ with flow capacities $r_j$ and transportation costs along the arcs $S_j$. Let us denote by $R\,[i_1, i_2]$ the set of paths from vertex $i_1$ to vertex $i_2$, let $D\,(\Pi_\alpha)$ be the set of arcs forming a path $\Pi_\alpha$, and let $K_j$ denote the set of paths passing through an arc $d_j$.

Suppose we have $p$ commodities and let $M_k$ denote the set of vertices producing the $k$-th commodity, while $N_k$ denotes the set of vertices demanding the $k$-th commodity $(k = 1, \ldots, p)$. For the $k$-th commodity, we denote by $a_i^{(k)}$ production capacities of vertices $i \in M_k$, and by $b_i^{(k)}$ demands of vertices $i \in N_k$. We shall assume that for all $k$

$$\sum_{i \in M_k} a_i^{(k)} = \sum_{i \in N_k} b_i^{(k)}.$$

Let us define the cost of transportation along the path $\Pi_\alpha$

$$c(\Pi_\alpha) = \sum_{j \in D(\Pi_\alpha)} S_j. \tag{4.30}$$

Denote by $x_{ii'}^{(k)}\,[\Pi_\alpha]\ (k = 1, \ldots, p;\ i \in M_k;\ i' \in N_k;\ \Pi_\alpha \in R\,[i, i'])$ the volume of the $k$-th product delivered from point $i$ to point $i'$ along the path $\Pi_\alpha$. Then the multicommodity transportation problem can be formulated as the problem of finding

$$\min \sum_{k=1}^{p} \sum_{i \in M_k} \sum_{i' \in N_k} \sum_{\Pi_\alpha \in R[i,i']} c\,(\Pi_\alpha)\, x_{ii'}^{(k)}\,[\Pi_\alpha], \tag{4.31}$$

subject to the constraints

$$x_{ii'}^{(k)}\,[\Pi_\alpha] \geqq 0, \tag{4.32}$$

$$\sum_{i' \in N_k} \sum_{\Pi_\alpha \in R[i,i']} x_{ii'}^{(k)}\,[\Pi_\alpha] = a_i^{(k)}, \tag{4.33}$$

$$\sum_{i \in M_k} \sum_{\Pi_\alpha \in R[i,i']} x_{ii'}^{(k)}\,[\Pi_\alpha] = b_i^{(k)}, \tag{4.34}$$

$$\sum_{k=1}^{p} \sum_{i \in M_k} \sum_{i' \in N_k} \sum_{\Pi_\alpha \in K_j} x_{ii'}^{(k)}\,[\Pi_\alpha] \leqq r_j. \tag{4.35}$$

The dual problem is to find

$$\max \left[ \sum_{k=1}^{p} \left( \sum_{i' \in N_k} b_{i'}^{(k)}\, v_{i'}^{(k)} - \sum_{i \in M_k} a_i^{(k)}\, U_i^{(k)} - \sum_j r_j\, W_j \right) \right], \tag{4.36}$$

subject to

$$W_j \geqq 0; \tag{4.37}$$

$$v_i^{(k)} - U_i^{(k)} - \sum_{j \in D[\Pi_\alpha]} W_j \leqq c\,[\Pi_\alpha]. \tag{4.38}$$

As in Example 1, we shall divide the variables of problem $(4.36)-(4.38)$ into two groups: the first group consisting of the variables $\{U_i^{(k)}\}$ and $\{W_j\}$, and the second one comprising $\{v_{i'}^{(k)}\}$. If we fix $U_i^{(k)} = \bar{U}_i^{(k)}$, $W_j = \bar{W}_j$ and consider problem $(4.36)-(4.38)$ with respect to $v_{i'}^{(k)}$, we obtain the trivial solution

$$v_{i'}^{(k)} = \min_{i,\, \Pi_\alpha \in R[i,\, i']} \left[ \bar{U}_i^{(k)} + \sum_{j \in D[\Pi_\alpha]} \bar{W}_j + c\,[\Pi_\alpha] \right]$$

$$= \min_{i,\, \Pi_\alpha \in R[i,\, i']} \left[ \bar{U}_i^{(k)} + \sum_{j \in D[\Pi_\alpha]} (S_i + \bar{W}_j) \right]. \tag{4.39}$$

It is also easy to determine the subgradient vector $g(\{\bar{U}_i^{(k)}\}, \{\bar{W}_j\})$. The component of this vector $\Delta_{i*}^{(k)}$, corresponding to $U_{i*}^{(k)}$, equals $\sum_{i' \in T_{i*}} b_{i'}^{(k)} - d_{i*}^{(k)}$, where the set $T_{i*}$ consists of indices $i'$ such that the minimum in (4.39) is attained for $i = i*$ (if the minimum is attained for many values of $i$, then any of these values may serve as $i*$).

The component of the vector $g$ corresponding to $W_j$ is computed by the formula

$$\Delta_j = \sum_{(i',\, k) \in L_j} b_{i'}^{(k)} - r_j,$$

where $L_j$ consists of the pairs $(i', k)$ for which the minimum in (4.39) is attained at $\Pi_\alpha \in K_j$.

In this way, by using the general decomposition scheme, we have constructed an algorithm for solving the dual of the multicommodity transportation problem. The solution of a particular version of this problem, when each product is delivered by only one supplier, was considered in detail in [92], where the subgradient method was used, too.

**Decomposition with Respect to Constraints**

The scheme of decomposition with respect to constraints is, in a sense, dual to the schemes of decomposition with respect to variables.

Let us consider the following convex programming problem with constraints divided into two groups: find

$$\min f_0(x), \tag{4.40}$$

subject to the constraints

$$f_i(x) \leq 0, \quad i = 1, \ldots, m, \tag{4.41}$$

$$\varphi_j(x) \leq 0, \quad j = 1, \ldots, n. \tag{4.42}$$

Consider the problem of finding

$$L(u) = \min_{x \in D} \left[ f_0(x) + \sum_{i=1}^{m} u_i f_i(x) \right],$$

where $u = \{u_i\}_{i=1}^m$ is an $m$-vector with nonnegative components, and $D$ is the subset of $E_l$ defined by constraints (4.42).

The following assertion is a consequence of the Kuhn-Tucker theorem: if a solution to problem (4.40)−(4.42) exists and constraints (4.41) and (4.42) satisfy the Slater constraint qualification and the set $D$ is bounded, then $L$ is a concave function defined for all $u \geqq 0$, the maximum of $L$ exists and is attained at a point $u^*$, which is the $u$-component of the saddle point of the Lagrange function of problem (4.40)−(4.42):

$$\Phi(x, u, v) = f_0(x) + \sum_{i=1}^m u_i f_i(x) + \sum_{j=1}^n v_j \varphi_j(x).$$

Searching for the maximum of the concave function $L$ is equivalent to searching for the minimum of the convex function $L_1(u) = -L(u)$. The minimum of the convex function $L_1$, subject to the constraint $u \geqq 0$, may be found by using the subgradient algorithm with projection on the set $\{u : u \geqq 0\}$. We remark that the operation of projection on this set is very simple: the nonnegative coordinates of the projected vector are left unchanged, while the negative ones are set to zero. For a special case of linear programming problems, the presented scheme of decomposition was considered in detail in [77].

## 4.2  An Iterative Method for Solving Linear Programming Problems of Special Structure

The main idea of the iterative algorithm given below consists in an application of a special decomposition scheme and the use of the subgradient algorithm for determining parameters of the subproblems into which the original problem is decomposed. In the case of the classical problem of optimal production planning [47], the proposed method has a simple interpretation. The set of technological modes of production is divided into groups $T_1, \ldots, T_l$, and for each group $T_k$ a vector of resource constraints $b^{(k)}$ is chosen so that $\sum_{k=1}^l b^{(k)} = b$, where $b$ is the common resource limit. For each group $T_k$ a partial optimization problem is solved and the solution is used for determining a direction of change of the vectors $b^{(k)}$.

The method will be described for linear programming problems of the form: find

$$\max(c, x) \tag{4.43}$$

subject to the constraints

$$A x \leqq b, \tag{4.44}$$

$$x \geqq 0 \tag{4.45}$$

under the assumption that the entries of the matrix $A$ and the components of the vectors $b$ and $c$ are nonnegative.

Let us write problem $(4.43)-(4.45)$ in the following form: maximize

$$L(x) = (c, x) = \sum_{j=1}^{n} c_j x_j, \tag{4.46}$$

subject to

$$\sum_{j=1}^{n} a_{ij} \leqq b_i, \quad i = 1, \ldots, m, \tag{4.47}$$

$$x_j \geqq 0, \quad j = 1, \ldots, n. \tag{4.48}$$

We assume that

$$c_j \geqq 0, \quad b_i \geqq 0, \quad a_{ij} \geqq 0, \quad i = 1, \ldots, m, \quad j = 1, \ldots, n. \tag{4.49}$$

We shall divide the set of indices $N = \{1, \ldots, n\}$ into $l$ disjoint nonempty subsets $N_t$, $1 \leqq t \leqq l$, and consider the partial problems: maximize

$$L_t(x^{(t)}) = \sum_{j \in N_t} c_j x_j, \tag{4.50}$$

subject to

$$\sum_{j \in N_t} a_{ij} x_j \leqq b_{it}, \quad i = 1, \ldots, m, \tag{4.51}$$

$$x^{(t)} \geqq 0, \tag{4.52}$$

where $x^{(t)} = \{x_j\}$, $j \in N_t$, $t = 1, \ldots, l$. If $b_{it} \geqq 0$, $i = 1, \ldots, m$, $t = 1, \ldots, l$, then each subproblem $(4.50)-(4.52)$ has a solution. We shall also assume that each column of the matrix $A$ has at least one nonzero element, which ensures the boundedness of the solution sets of the partial problems. If

$$\sum_{k=1}^{l} b_{ik} = b_i, \quad i = 1, \ldots, m, \tag{4.53}$$

$$b_{ik} \geqq 0, \quad k = 1, \ldots, l, \tag{4.54}$$

then the vector $x \{b_{ik}\}$, formed from the solutions of the partial problems $(4.50)-(4.53)$, is feasible for problem $(4.46)-(4.48)$ and

$$L(x \{b_{ik}\}) = \sum_{k=1}^{l} L_k(\bar{x}^{(k)}),$$

where $\bar{x}^{(k)}$ is the solution of $(4.50)-(4.52)$. Let $\{u_{ik} \{b_{ik}\}\}$, $i = 1, \ldots, m$, denote a solution to the dual of $(4.46)-(4.48)$. Then, by the duality theorem,

$$\Phi \{b_{ik}\} = L(x \{b_{ik}\}) = \sum_{k=1}^{l} \sum_{i=1}^{m} b_{ik} \bar{u}_{ik} \{b_{ik}\}. \tag{4.55}$$

The function $F = -\Phi$ is a convex, piecewise linear function. Its subgradient at $\{b_{ik}\}$ may be computed by the formula

$$g_F\{b_{ik}\} = -g_\Phi\{b_{ik}\} = \{-\bar{u}_{ik}\{b_{ik}\}\}.$$

In this way, problem (4.46)–(4.48) is reduced to the problem of minimizing the convex function $F$ subject to constraints (4.53) and (4.54). Let $D$ denote the feasible set defined by those constraints. It is easy to observe that $D$ is a closed, convex polyhedron. Consider a sequence of vectors $\beta_p = \{b_{ik}\}_p$ generated by the formula

$$\beta_{p+1} = P_D\{\beta_p + h_{p+1}\,g_\Phi(\beta_p)\} = P_D\{\beta_p + h_{p+1}\,u_p\}, \tag{4.56}$$

where $u_p = \{\bar{u}_{ik}\{b_{ik}\}\}_p$, $\{h_p\}_{p=0}^\infty$ is a sequence of positive numbers, and $P_D$ is the operator of projection on $D$. Clearly, formula (4.56) describes one iteration of the subgradient algorithm with projection on the feasible set for minimizing the function $F(\beta) = F\{b_{ik}\}$. By Polyak's theorem [64], if $h_p \to 0$ and $\sum_{p=0}^\infty h_p = +\infty$ then $F(\beta_p)$ converges to the minimum.

We shall now consider the sequence $x(\{b_{ik}\}_p)$. It follows from (4.55) that $L(x\{b_{ik}\}_p)$ converges to the optimal value of problem (4.46)–(4.48). Let us describe the projection operation involved in (4.56) in a more detailed way. It consists in finding for any point $\beta$ its nearest point $P_d(\beta)$ in the convex polyhedron $D$, which can be expressed as a Cartesian product of simplices:

$$D = D_1 \times D_2 \times \ldots \times D_m,$$

where each simplex $D_i$ is defined by the constraints

$$\sum_{k=1}^l y_{ik} = b_i, \tag{4.57}$$

$$y_{ik} \geq 0, \quad k = 1, \ldots, l. \tag{4.58}$$

Let $\beta_i = \{b_{ik}\}$, $k = 1, \ldots, l$. Then the projection operation $P_D$ reduces to independent projections of vectors $\beta_i$ on the simplices $D_i$ for all $i = 1, \ldots, m$. We obtain the following elementary quadratic programming problem of minimizing

$$\varrho^2(\{y_{ik}\}, \beta_i) = \sum_{k=1}^l (y_{ik} - b_{ik})^2, \tag{4.59}$$

subject to constraints (4.57) and (4.58).

Let us consider the family of problems, depending on a parameter $\lambda$: find

$$\min\left[\sum_{k=1}^l (y_{ik} - b_{ik})^2 + 2\lambda \sum_{k=1}^l y_{ik}\right], \tag{4.60}$$

subject to $y_{ik} \geq 0$.

For any $\lambda$ this problem has the following easily verifiable solution

$$y_{ik}(\lambda) = \max(0, b_{ik} - \lambda), \quad k = 1, \ldots, l. \tag{4.61}$$

By the Kuhn-Tucker theorem, there exists $\lambda^* = \lambda$ for which the solution of (4.60) coincides with the solution of problem $(4.57) - (4.59)$. Such $\lambda^*$ can be found in a finite number of iterations by the following procedure. Let

$$\lambda_1 = \frac{\sum\limits_{k=1}^{l} b_{ik} - b_i}{l} \, .$$

Next, for any $q$, $q = 1, 2, \ldots$, we proceed as follows. Suppose we have $\lambda_q$. Let us compute $y_{ik}(\lambda_q)$ by formula (4.61). Set

$$S_q = \sum_{k=1}^{l} y_{ik}(\lambda_q) \, .$$

If $S_q = b_i$, we set $\lambda^* = \lambda_q$. If $S_q > b_i$ then we set

$$\lambda_{q+1} = \lambda_q + \frac{S_q - b_i}{n_q} \, , \tag{4.62}$$

where $n_q$ is the number of indices $k$ for which $y_{ik}(\lambda_q) > 0$, and we proceed to the next iteration. It follows from formulae (4.61) and (4.62) that the case $S_q < b_i$ cannot occur. Moreover, $l > n_1 > n_2 > \ldots$ . Therefore, the process of choosing $\lambda^*$ must terminate after a finite number of iterations. Since $\lambda^*$ is the unique value of $\lambda$ for which $\{y_{ik}(\lambda)\}$ satisfies constraints $(4.57) - (4.58)$, $\{y_{ik}(\lambda^*)\}$ is a solution of problem $(4.57) - (4.59)$, i.e. $\{y_{ik}(\lambda^*)\}$ is the projection of $\beta_i$ onto $D_i$.

Thus at each iteration of the algorithm for solving problem $(4.43) - (4.45)$ according to the above-described decomposition scheme we have to perform the following operations:

1) for given $\{b_{ik}\}_p$ we solve $l$ partial problems $(4.50) - (4.52)$ and find shadow prices $\{u_{ik}\}$;
2) we compute $\{b_{ik}\}_{p+1}$ by using formula (4.56).

Of course, such a decomposition technique is useful if each partial problem is simple, for instance if $l$ equals $n$, i.e. when we have $n$ partial problems, but each contains only one unknown. In this case the $j$-th component of the vector $x(\{b_{ik}\})$ can be computed by the formula

$$x_j(\{b_{ik}\}) = \min_{i \in I_j^+} \frac{b_{ij}}{a_{ij}} \, , \tag{4.63}$$

where $I_j^+$ is the set of $i$-s with $a_{ij} > 0$.

Let $i_j$ be an index at which the minimum in (4.63) is attained. Then

$$u_{ij} = \begin{cases} c_j/a_{ij} & \text{for } i = i_j, \\ 0 & \text{for } i \neq i_j. \end{cases} \tag{4.64}$$

We also observe that in this case the vector $g_\Phi(\{b_{ik}\})$ has at most $n$ nonnegative components.

Formulae (4.63) and (4.64) enable one to calculate the subgradient $g_\Phi(\{b_{ij}\})$ in an extremely simple way. Using these formulae, one may easily design an iterative algorithm for solving linear programming problems of the form given above [91].

## 4.3 The Solution of Distribution Problems by the Subgradient Method

Distribution problems are, in our terminology, linear programming problems of the following form:
find

$$\min \sum_{i,\alpha} c_{i\alpha} x_{i\alpha}, \tag{4.65}$$

subject to

$$\sum_{\alpha \in M_i} x_{i\alpha} = b_i, \quad i \in I = \{1, \ldots, n\}, \tag{4.66}$$

$$\sum_{i \in I; \alpha \in M_\mu} p_{i\alpha}^{(\mu)} x_{i\alpha} \leq a^{(\mu)}, \quad \mu \in T, \tag{4.67}$$

$$x_{i\alpha} \geq 0. \tag{4.68}$$

The set $I$ will be called the set of demands and the set $T$ will be called the set of technological constraints. $M_i$ and $M_\mu$ are subsets of some finite set $M$. Problems of the form $(4.65)-(4.68)$ arise very frequently in practice, especially in the solution of industrial planning problems. Particular cases include the usual transportation and allocation problems ($\lambda$-problems or $k$-problems). The research on gradient methods for solving distribution problems stems from the necessity to solve real-life, large scale problems connected with planning optimum loads of turning lathes of the USSR and distributing the production among customers, the size of the problems and the necessity to use medium capacity computers (of the type M-220 and "Minsk-32") dictating definite requirements on the algorithm, especially with respect to the use of the memory of a computer. The subgradient method turned out to be applicable to the solution of such problems.

The dual of problem $(4.65)-(4.68)$ is formulated as follows:
find

$$\max \left( \sum_{i \in I} b_i v_i - \sum_{\mu \in T} a^{(\mu)} u_\mu \right), \tag{4.69}$$

subject to

$$v_i - \sum_\mu p_{i\alpha}^{(\mu)} u_\mu \leq c_{i\alpha}, \quad \alpha \in M_\mu, \quad i \in I, \tag{4.70}$$

$$u_\mu \geq 0. \tag{4.71}$$

By (4.66), problem (4.65)−(4.68) has a solution if $b_i \geqq 0$. Under this assumption, we shall apply to the dual problem (4.69)−(4.71) the scheme of decomposition with respect to variables, assigning the variables $\{v_i\}$ to the first group, and the variables $\{u_\mu\}$ to the second. Fixing $\{u_\mu\}$, we obtain the following problem with respect to the variables $\{v_i\}$:

maximize

$$\sum_i b_i v_i, \tag{4.72}$$

subject to

$$v_i \leqq c_{i\alpha} + \sum_\mu p_{i\alpha}^{(\mu)} u_\mu. \tag{4.73}$$

With $b_i \geqq 0$, this problem has the trivial solution

$$v_i = \min_\alpha \left( c_{i\alpha} + \sum_\mu p_{i\alpha}^{(\mu)} u_\mu \right). \tag{4.74}$$

Using the results of Sect. 4.1, we obtain the following algorithm for solving the dual of the distribution problem: we choose initial values $u_\mu^0 \geqq 0$ and after $k$ iterations obtain a vector $\{u_\mu^{(k)}\}$. Then the $(k+1)$-st iteration consists of the following steps:

(1) we calculate $v_i^{(k)}$ by formula (4.74), setting $u_\mu = u_\mu^k$, and find $\alpha(k)$ as one of the values of $\alpha$ at which the minimum in (4.74) is attained;

(2) we set

$$x_{i\alpha}^{(k)} = \begin{cases} b_i & \text{if } \alpha = \alpha(k), \\ 0 & \text{if } \alpha \neq \alpha(k); \end{cases}$$

(3) we calculate the residuals

$$\Delta_\mu^{(k)} = a_\mu - \sum_\mu p_{i\alpha}^{(\mu)} x_{i\alpha};$$

(4) we compute the next approximation to $\{u_\mu\}$

$$u_\mu^{k+1} = \max \{0, u_\mu - h_{k+1} \Delta_\mu^{(k)}\}.$$

The choice of the stepsizes $\{h_k\}$, ensuring convergence to an optimal solution, if it exists, was discussed in Sect. 2.2.

To sum up, the scheme of decomposition with respect to variables combined with the use of the subgradient method yields a very simple algorithm for solving the dual problem (4.69)−(4.71).

It is not a trivial task in this case to recover the solution to the primal (4.65)−(4.68) from the approximate solution to the dual. The linear programming duality theory suggests the following algorithm for selecting the variables $\{x_{i\alpha}\}$ that enter the optimal basis of the primal problem: one has to choose a sufficiently small number $\delta > 0$ and select those pairs of indices $(i, \alpha)$, for which the following inequality is satisfied

$$c_{i\alpha} + \sum_{\mu} p_{i\alpha}^{(\mu)} \bar{u}_{\mu} - \delta \leq \bar{v}_i, \tag{4.75}$$

where $\{\bar{u}_{\mu}, \bar{v}_i\}$ are the estimates of the dual variables obtained by the above-described iterative algorithm for solving problem $(4.69)-(4.71)$. Having found the set of optimal basic variables $\{x_{i\alpha}\}$, it is easy to obtain the optimal value of $\{x_{ia}\}$ by solving the corresponding system of linear algebraic equations.

However, such a procedure is not always correct. If the solution to problem $(4.65)-(4.69)$ is nonunique or there are several bases with the objective function values close to the optimum, then the choice of basic variables by inequality (4.75) may yield spurious variables and our attempt to recover the solution to the primal problem may fail. Moreover, even if we obtain the required basis, for large scale problems the solution of the corresponding set of linear equations needs the use of either a slow external memory or special procedures. Therefore, one had to find a simpler and reliable procedure for obtaining the solution to the primal problem. Such a procedure was designed on the basis of the idea of "smoothing", which was applied, with no formal substantiation, by Gerstenhaber in the construction of a procedure for solving the transportation problem [36].

Let us construct the function $S(u_{\mu})$ by substituting for $v_i$ in the expression of the objective function (4.69) the right side of formula (4.74):

$$S(\{u_{\mu}\}) = \sum_{i \in I} b_i \min_{\alpha} \left( c_{i\alpha} + \sum_{\mu} p_{i\alpha}^{(\mu)} u_{\mu} \right) - \sum_{\mu} a^{(\mu)} u_{\mu}.$$

The function $S$ is piecewise linear and convex. A subgradient of this function at $u = \{u_{\mu}\}$ can be calculated by the formula

$$g_S(\{u_{\mu}\}) = \left\{ \sum_{i} b_i p_{i\alpha*}^{(\mu)} - a^{(\mu)} \right\},$$

where $\alpha*$ is one of the values of $\alpha$ at which the minimum in (4.74) is attained.

Let us consider the vector field $G = \{g_S(u)\}$, $u = \{u_{\mu}\}$. It is not continuous: its points of discontinuity correspond to those vectors $u$, for which the minimum in formula (4.74) is nonunique.

Having selected $\delta > 0$, we shall construct a "$\delta$-smoothed" vector field $G_{\delta}$, defined at point $u$ by the vector $g^{\delta}(u)$ according to the following formulae:

$$g^{\delta}(u) = \left\{ \sum_{i,\alpha} p_{i\alpha}^{(\mu)} b_i m_{i\alpha}^{(\delta)}(u) - a^{(\mu)} \right\},$$

where

$$m_{i\alpha}^{(\delta)} = \frac{\Delta_{i\alpha}^{(\delta)}}{\sum_{j} \Delta_{j\alpha}^{(\delta)}},$$

and $\Delta_{i\alpha}^{(\delta)}$ is calculated by the formulae

$$\Delta_{i\alpha}^{(\delta)} = \begin{cases} 0 & \text{if } r_{i\alpha}^{(\delta)} = \left(c_{i\alpha} + \sum_{\mu} p_{i\alpha}^{(\mu)} u_{\mu}\right) - \left(c_{i\alpha*} + \sum_{\mu} p_{i\alpha*}^{(\mu)} u_{\mu}\right) \leq \delta, \\ \delta - r_{i\alpha}^{(\delta)} & \text{otherwise.} \end{cases}$$

To take account of the constraints $u \geq 0$ we shall consider the vector field $G_r^{\delta}$ defined by vectors of the form

$$g_r^{\delta}(u) = g^{\delta}(u) + r\,u^+,$$

where $u^+$ is the vector obtained from the vector $u$ by setting its negative components to zero, and $r$ is some small positive number.

**Lemma 4.1.** *The vector field $G_r$ is continuous for any $\delta > 0$ and $r > 0$.*

*Proof.* The continuity of the vector field $g_r^{\delta}(u)$ follows from the continuity of the functions $\Delta_{i\alpha}^{(\delta)}$ and $m_{i\alpha}^{(\delta)}$ by which it is defined. The continuity of the vector field $r\,u^+$ may be also easily verified. $\square$

**Lemma 4.2.** *The continuous vector field $G_r^{\delta}$ for any $\delta > 0$ and $r > 0$ has at least one fixed point, provided that problem* (4.65)–(4.68) *has a unique solution.*

*Proof.* Consider a straight line passing through the optimal point $x^*$ of problem (4.65)–(4.68). Let $x_R^{(1)}$ and $x_R^{(2)}$ be the intersection points of this straight line with the ball of radius $R$ centered at $x^*$. It is easy to show that for sufficiently large $R$ the field $G_r^{\delta}$ forms at the points $x_R^{(1)}$ and $x_R^{(2)}$ vectors $g_1^{(R)}$ and $g_2^{(R)}$ satisfying $(g_1^{(R)}, x_R^{(1)} - x^*) > 0$ and $(g_2^{(R)}, x_R^{(2)} - x^*) > 0$. It follows that the vectors $g_1^{(R)}$ and $g_2^{(R)}$ are not colinear. By using a corollary of Borsuk's theorem on the circulation of continuous vector fields [49], we deduce the existence of a fixed point of $G_r^{\delta}$ for each $r > 0$ and $\delta > 0$. $\square$

Passing to the limit with $r \to \infty$, we deduce for each $\delta > 0$ the existence of a point $u^{(\delta)*}$ satisfying

$$\sum_{i,\alpha} p_{i\alpha}^{(\mu)} b_i m_{i\alpha}^{(\delta)} \{u_{\mu}^{(\delta)*}\} - a^{(\mu)} = 0, \quad u_{\mu}^{(\delta)*} > 0,$$

$$\sum_{i,\alpha} p_{i\alpha}^{(\mu)} b_i m_{i\alpha}^{(\delta)} \{u_{\mu}^{(\delta)*}\} - a^{(\mu)} \leq 0, \quad u_{\mu}^{(\delta)*} = 0. \tag{4.76}$$

If we let

$$\bar{x}_{i\alpha}^{(\delta)} = b_i\, m_{i\alpha}^{(\delta)} (u^{(\delta)*}),$$

then conditions (4.76) show that $\bar{x}_{i\alpha}^{(\delta)}$ is feasible for problem (4.65)–(4.68). Since the objective value at any feasible point of the primal problem is not smaller than the dual objective value at any feasible point of the dual, we obtain the inequality

$$\sum_{i,\alpha} c_{i\alpha}\, \bar{x}_{i\alpha}^{(\delta)} \geq \sum_{i \in I} b_i \min_{\alpha} \left(c_{i\alpha} + \sum_{\mu} p_{i\alpha}^{(\mu)} u_{\mu}^{(\delta)*}\right) - \sum_{\mu} a^{(\mu)} u_{\mu}^{(\delta)*}.$$

With $\delta \to 0$, the left and right sides of the above inequality converge to each other, and hence to the optimal value.

One approach to obtaining the primal solution is based on the following observations. The number of nonzero $x_{i\alpha}$ in the basic solution of problem $(4.65)-(4.68)$ does not exceed $n+m$. In large problems that arise in applications (mainly such problems are solved by iterative methods), one has $n \gg m$, with the value of $n/m$ being in the tens, sometimes larger than a hundred. This means that in the optimal basis of problem $(4.65)-(4.68)$ the number of indices $i$ for which the number of nonzero $x_{i\alpha}$, $\alpha \in M_i$, exceeds unity is no larger than $m$. We also note that if $\{v_i^*, u_\mu^*\}$ is the dual solution and problem $(4.65)-(4.68)$ admits of a unique solution, then the equality

$$v_i^* = \min_\alpha \left\{ c_{i\alpha} + \sum_\mu p_{i\alpha}^{(\mu)} u_\mu^* \right\} \tag{4.77}$$

is attained for $\alpha$-s that correspond to $x_{i\alpha} > 0$ in the primal solution. Consequently, if we have managed to obtain sufficiently accurate approximations to the optimal values $\{u_\mu\}$, then by selecting $\varepsilon > 0$ depending on the attained accuracy of $\{u_\mu\}$ and by using equality (4.76), we obtain the following approximate criterion for testing if $x_{i\alpha} > 0$ in the solution to problem $(4.65)-(4.68)$:

$$v_i - \varepsilon \leqq \min_\alpha \left\{ c_{i\alpha} + \sum_\mu p_{i\alpha}^{(\mu)} u_\mu \right\}. \tag{4.78}$$

Let $\bar{I}$ denote the set of indices $i$ for which inequality (4.78) holds for one $\alpha = \alpha(i)$. First, we set $x_{i\alpha(i)} = b_i$ and $x_{i\alpha} = 0$ if $\alpha \neq \alpha(i)$, $i \in \bar{I}$. Then, in a natural way we obtain a distribution problem with the set of demands $I_1 = I \backslash \bar{I}$ and the sets $M_i$ consisting of those indices $\alpha$ for which inequality (4.78) is satisfied. From the above considerations it is clear that in problems of practical interest the number of elements of $I_1$ is less by an order than the number of elements of $I$. In this way we arrive at the problem of the following form:
find

$$\min \sum c_{i\alpha} x_{i\alpha}, \tag{4.79}$$

subject to

$$\sum_{\alpha \in M_i} x_{i\alpha} = b_i, \quad i \in I_1, \tag{4.80}$$

$$\sum_{i \in I_1} \sum_{\alpha \in M_\mu} p_{i\alpha}^{(\mu)} x_{i\alpha} \leqq a^{(\mu)} - \sum_{i \in \bar{I}} b_i p_{i\alpha}^{(\mu)}, \tag{4.81}$$

$$x_{i\alpha} \geqq 0. \tag{4.82}$$

This problem of a relatively small size can be solved by using either some finite algorithm, which is efficient for problems of such dimensions, or the algorithm with "smoothing". These approaches need no further comments. We only note that since problem $(4.79)-(4.82)$ has a significantly smaller size than problem $(4.65)-(4.68)$, in practice its solution uses $5-10\%$ of the time required for solving the original problem.

When solving numerous practical distribution problems by the subgradient method in the space of the dual variables, we recovered solutions to the primal problems by the algorithm with "smoothing".

A standard program, implemented by G. I. Gorbach [90] for computers of the type M-20, was used for solving various distribution problems, in particular large scale transportation problems and problems of allocating ships to river transportation lines. Distribution problems of particularly large scale were solved in planning loads of turning lathes [90]. In those problems the number of demands reached 10,000, the number of lathes exceeded 50 and the range of products comprised about 1000 items. The algorithms proposed were used for planning the load of lathes in "Soyuzglavmetall" of Gossnab USSR with a significant econimic effect.


## 4.4  Experience in Solving Production-Transportation Problems by Subgradient Algorithms with Space Dilation

We shall consider several mathematical models of operative planning, which can be in some way converted to the problem of minimizing a nonsmooth convex function subject to simple constraints. Up till now the researchers working at the Institute of Cybernetics of the Academy of Sciences of the Ukrainian Soviet Socialist Republic (N. Z. Shor, N. G. Zhurbenko, V. I. Gershovich, T. B. Belykh, L. V. Belaeva and others) have gained considerable experience in solving operational planning problems by the $r$-algorithms, which can be quite efficiently applied in many cases. We shall consider several most typical examples.

### 1. The Dynamic Transportation Problem

There are $m$ suppliers (plants) and $n$ consumers (building sites). The planning interval will be called a quarter. The following quantities are given:

$a_{it}$ — the volume of the production of the $i$-th plant in the $t$-th quarter, $t = \overline{1, T}$;

$a_i$ — the volume of the production of the $i$-th plant in the planning period; $a_i = \sum_{t=1}^{T} a_{it}$;

$b_{jt}$ — the volume of the demand of the $j$-th building site in quarter $t$, $t = \overline{1, T}$;

$b_j$ — the demand of the $j$-th building site for supplies in the planning period, $b_j = \sum_{t=1}^{T} b_{jt}$;

$c_{ij}$ — the unit cost of transportation from the $i$-th supplier to the $j$-th consumer.

The transportation costs are assumed to be independent from the time of delivery. It is assumed that the total production and demand are balanced:

$$\sum_{i=1}^{m} a_i = \sum_{j=1}^{n} b_j,$$

but a quarterly balance, in general, need not occur:

$$\sum_{i=1}^{m} a_{it} \text{ is not necessarily equal to } \sum_{j=1}^{n} b_{jt}.$$

Intuitively, the problem consists in finding a distribution of products among the consumers so that the following conditions are satisfied:

(a) for most consumers, or at least for the consumers with a highest priority, their quarterly demands are met;
(b) the volume of products turned out ahead of time should be as small as possible (this condition may be interpreted as the minimization of the losses incurred in storing the foreward production);
(c) the transportation costs should be minimum with respect to other allocations that satisfy the first two conditions;
(d) the consumers should, if possible, stick to one supplier throughout the planning period.

Let us introduce the following notations:

$\beta_{jt}$ – the volume of the demand of the $j$-th consumer in the first $t$ quarters

$$\beta_{jt} = \sum_{\tau=1}^{t} b_{j\tau}, \quad t \in \overline{1, T};$$

$\alpha_{it}$ – the volume of the production of the $i$-th plant in the first $t$ quarters

$$\alpha_{it} = \sum_{\tau=1}^{t} a_{i\tau}, \quad t = \overline{1, T};$$

$x_{ijt}$ – the volume of the production supplied by the $i$-th plant to the $j$-th consumer in the first $t$ quarters.

We shall assume that all $\beta_{jt} \neq 0$. However, we note that this limitation, as we shall see in what follows, is not essential and is introduced only for simplicity of exposition. The final results are valid for the case $\beta_{jt} \geq 0$. The variable $x_{ijt}/\beta_{jt}$ will be called the degree of provision of the $j$-th consumer by the $i$-th supplier in the first $t$ quarters.

To construct a mathematical model we shall augment the statement of the problem, without changing significantly its economic substance, with the following conditions:

(1) the forward production of a plant is stored in its warehouses, i.e. the consumers do not receive the excessive production:

$$\sum_{i=1}^{M} x_{ijt} \leq \beta_{jt}, \quad t = \overline{1, T},$$

(2) the degree of provision of a consumer does not decrease in time:

$$\frac{x_{ijt}}{\beta_{jt}} \geqq \frac{x_{ij(t-1)}}{\beta_{j(t-1)}}, \quad t = \overline{2, T}. \tag{4.83}$$

Let us introduce the penalty coefficients: $R_{jt}$ denoting the penalty for not supplying a unit of products to the $j$-th consumer by the end of the $t$-th quarter, and $l_{it}$ denoting the penalty for the existence of a unit of undistributed products in the $i$-th plant at the end of the $t$-th quarter. By $w_{it}$ we shall denote the volume of undistributed products of the $i$-th enterprise at the end of the $t$-th quarter.

The mathematical model of the problem in question is given by the following linear programming problem:
minimize

$$\left[ \sum_{j,i=1}^{n,m} c_{ij} x_{ijT} + \sum_{j,t=1}^{n,T-1} R_{jt} \left( \beta_{jt} - \sum_{i=1}^{n} x_{ijt} \right) + \sum_{i,t=1}^{m,T-1} l_{it} w_{it} \right], \tag{4.84}$$

subject to

$$\sum_{i=1}^{m} x_{ijT} = b_j, \quad j = \overline{1, n}, \tag{4.85}$$

$$\sum_{j=1}^{n} x_{ijT} = a_i, \quad i = \overline{1, m}, \tag{4.86}$$

$$\beta_{jt-1} x_{ijt} \geqq \beta_{jt} x_{ijt-1}, \quad i = \overline{1, m}, \quad j = \overline{1, n}, \quad t = \overline{2, T} \tag{4.87}$$

$$\sum_{j=1}^{n} x_{ijt} + w_{it} = \alpha_{it}, \quad i = \overline{1, m}, \quad t = \overline{1, T-1}, \tag{4.88}$$

$$\sum_{i=1}^{m} x_{ijt} \leqq \beta_{jt}, \quad j = \overline{1, n}, \quad t = \overline{1, T-1}, \tag{4.89}$$

$$x_{ijt} \geqq 0, \quad t = \overline{1, T}, \quad i = \overline{1, m}, \quad j = \overline{1, n}, \tag{4.90}$$

$$w_{it} \geqq 0, \quad t = \overline{1, T-1}, \quad i = \overline{1, m}. \tag{4.91}$$

**Remarks**

1. Constraints (4.87), which make sense also for $\beta_{jt} = 0$, correspond to condition (4.83).
2. Although the natural constraint $x_{ijt+1} \geqq x_{ijt}$ has not been included in the list of the problem constraints, it is a consequence of constraints (4.87), (4.89) and (4.90), because $\beta_{jt+1} \geqq \beta_{jt}$.
3. We note that condition (4.87) forbids a plant to interrupt its supplying a given consumer, which aids in fulfilling condition (d). A more exact formulation of the requirement (d) would lead to an integer programming problem, the solution of which would be rather problematic if the problem is large.

We shall now describe a method for solving the problem in question. The solution of problem $(4.84)-(4.91)$ will proceed in two stages. On the first stage we determine the optimal assignment of the consumers to the suppliers, while on the second stage the volumes of the quarterly supplies $x_{ijt}$ are found. The search for an optimal assignment of the consumers reduces, as in the case of the classical transportation problem, to the solution of the dual problem.

**Remark.** By an optimal assignment we mean the selection of the pairs of indices $(i, j)$ for which $x_{ijT} > 0$.

Before passing to the dual problem we shall write problem $(4.84)-(4.91)$ in the following equivalent form:

minimize

$$\left[ \sum_{j,i=1}^{n,m} c_{ij} x_{ijT} + \sum_{j,\tau=1}^{n,T-1} \left( \sum_{t=1}^{\tau} \beta_{jt} R_{jt} \right) \left( \sum_{i=1}^{m} y_{ij\tau} \right) + \sum_{i,t=1}^{m,T-1} l_{it} w_{it} \right], \tag{4.92}$$

subject to

$$\sum_{i=1}^{m} x_{ijT} = b_j, \quad j = \overline{1, n}, \tag{4.93}$$

$$\sum_{j=1}^{n} x_{ijT} = a_i, \quad i = \overline{1, m}, \tag{4.94}$$

$$\sum_{j=1}^{n} \beta_j \sum_{\tau=0}^{t-1} y_{ij\tau} + w_{it} = \alpha_{it}, \quad i = \overline{1, m}, \quad t = \overline{1, T-1}, \tag{4.95}$$

$$-\frac{1}{b_j} x_{ijT} + \sum_{\tau=0}^{T-1} y_{ij\tau} = 0, \quad i = \overline{1, m}, \quad j = \overline{1, n}, \tag{4.96}$$

$$x_{ijT} \geqq 0, \quad i = \overline{1, m}, \quad j = \overline{1, n}, \tag{4.97}$$

$$y_{ij\tau} \geqq 0, \quad i = \overline{1, m}, \quad j = \overline{1, n}, \quad \tau = \overline{0, T-1}, \tag{4.98}$$

$$w_{it} \geqq 0, \quad i = \overline{1, m}, \quad t = \overline{1, T-1}. \tag{4.99}$$

It is not difficult to show that problem $(4.92)-(4.99)$ is equivalent to problem $(4.84)-(4.91)$, and that the solution $x_{ijt}$ to problem $(4.84)-(4.91)$ is given by the formula

$$x_{ijt} = \beta_{jt} \sum_{\tau=0}^{t-1} y_{ij\tau}, \quad i = \overline{1, m}, \quad j = \overline{1, n},$$

where $\{y_{ijt}\}$ is the solution to problem $(4.92)-(4.99)$.

We shall demonstrate, for instance, the truth of the equality

$$\sum_{\tau=1}^{T-1} \left( \sum_{t=1}^{\tau} \beta_{jt} R_{jt} \right) \left( \sum_{i=1}^{m} y_{ij\tau} \right) = \sum_{\tau=1}^{T-1} R_{j\tau} (\beta_{j\tau} - \tilde{x}_{j\tau}),$$

where

$$\tilde{x}_{j\tau} = \sum_{i=1}^{m} x_{ij\tau}.$$

Setting

$$r_{j\tau} = \sum_{t=1}^{\tau} \beta_{jt} R_{jt}, \quad \tau = \overline{1, T-1}, \quad j = \overline{1, m},$$

and taking account of (4.96), we obtain

$$\sum_{\tau=1}^{T-1} \left( \sum_{t=1}^{\tau} \beta_{jt} R_{jt} \right) \left( \sum_{i=1}^{m} y_{ij\tau} \right)$$

$$= \sum_{\tau=1}^{T-1} r_{j\tau} \sum_{i=1}^{m} \left( \frac{x_{ij\tau+1}}{\beta_{j\tau+1}} - \frac{x_{ij\tau}}{\beta_{j\tau}} \right)$$

$$= \sum_{\tau=1}^{T-1} r_{j\tau} \left( \frac{\tilde{x}_{j\tau+1} - \beta_{j\tau+1}}{\beta_{j\tau+1}} - \frac{\tilde{x}_{j\tau} - \beta_{j\tau}}{\beta_{j\tau}} \right)$$

$$= \sum_{\tau=1}^{T-1} r_{j\tau} \left( \frac{\beta_{j\tau} \tilde{x}_{j\tau}}{\beta_{j\tau}} \right) - \sum_{\tau=1}^{T-1} r_{j\tau} \left( \frac{\beta_{j\tau+1} - \tilde{x}_{j\tau+1}}{\beta_{j\tau+1}} \right)$$

$$= \sum_{\tau=1}^{T-1} r_{j\tau} \left( \frac{\beta_{j\tau} - \tilde{x}_{j\tau}}{\beta_{j\tau}} \right) - \sum_{\tau=1}^{T-1} (r_{j\tau+1} - \beta_{j\tau+1} R_{j\tau+1}) \frac{\beta_{j\tau+1} - \tilde{x}_{j\tau+1}}{\beta_{j\tau+1}}$$

$$= r_{j1} \frac{\beta_{j1} - \tilde{x}_{j1}}{\beta_{j1}} + \sum_{\tau=2}^{T-1} R_{j\tau} (\beta_{j\tau} - \tilde{x}_{j\tau})$$

$$= \sum_{\tau=1}^{T-1} R_{j\tau} (\beta_{j\tau} - \tilde{x}_{j\tau}),$$

as required.

Denote by $v_j, u_{iT}, u_{it}$ and $z_{ij}$ the dual variables corresponding to constraints (4.93)−(4.96), respectively. Then the dual problem (4.92)−(4.99) is the following:

maximize

$$\left( \sum_{j=1}^{n} b_j v_j - \sum_{i,t=1}^{m,T} \alpha_{it} u_{it} \right), \tag{4.100}$$

subject to

$$v_j \leq c_{ij} + \frac{1}{b_j} z_{ij} + u_{iT}, \quad i = \overline{1, m}, \quad j = \overline{1, n}, \tag{4.101}$$

$$z_{ij} \leq \sum_{t=1}^{T-1} \beta_{jt} u_{it}, \quad i = \overline{1, m}, \quad j = \overline{1, n}, \tag{4.102}$$

$$z_{ij} \leq \sum_{t=1}^{\tau} \beta_{jt} R_{jt} + \sum_{t=\tau+1}^{T-1} \beta_{jt} u_{it}, \quad i = \overline{1, m}, \quad j = \overline{1, n}, \quad \tau = \overline{1, T-2}, \quad (4.103)$$

$$z_{ij} \leq \sum_{t=1}^{T-1} \beta_{jt} R_{jt}, \quad i = \overline{1, m}, \quad j = \overline{1, n}, \quad (4.104)$$

$$u_{it} \geq - l_{it}, \quad i = \overline{1, m}, \quad t = \overline{1, T-1}. \quad (4.105)$$

Let $v_j^*$, $u_{it}^*$ and $z_{ij}^*$ denote a solution to problem $(4.100) - (4.105)$.

Then, clearly, we have the relation

$$v_j^* = \min_{i = \overline{1, m}} \left[ c_{ij} + \frac{1}{b_j} z_{ij}^* + u_{iT}^* \right]. \quad (4.106)$$

Let $I^*(j)$ denote the set of all indices $i$ at which the minimum in the right side of formula (4.106) is attained.

Since the variables $x_{ijT}$ of problem $(4.92) - (4.99)$ are shadow prices for constraints (4.101), $x_{ijT} > 0$ only if $i \in I^*(j)$. Therefore $I^*(j)$ is the set of suppliers to which the $j$-th consumer is assigned. Thus, to find an optimal assignment of consumers to producers, it suffices to solve the dual problem $(4.100) - (4.105)$. We note that if $R_{jt} = 0$ and $l_{it} = 0$, $i = \overline{1, m}$, $j = \overline{1, n}$, $t = \overline{1, T-1}$, then $(4.100) - (4.105)$ is the dual of the classical transportation problem, defined by the parameters $c_{ij}$, $a_i$ and $b_j$.

For solving problem $(4.100) - (4.105)$ we shall use the scheme of decomposition with respect to variables. Fixing the variables $u_{it}$, we arrive at the problem
maximize

$$\sum_{j=1}^{n} b_j v_j,$$

subject to

$$v_i \leq c_{ij} + \frac{1}{b_j} z_{ij} + u_{iT}, \quad i = \overline{1, m}, \quad j = \overline{1, n}, \quad (4.107)$$

$$z_{ij} \leq r_{ij\tau}(u), \quad i = \overline{1, m}, \quad j = \overline{1, n}, \quad \tau = \overline{0, T-1},$$

where

$$r_{ij0}(u) = \sum_{t=1}^{T-1} \beta_{jt} u_{it},$$

$$r_{ij\tau}(u) = \sum_{t=1}^{\tau} \beta_{jt} R_{jt} + \sum_{t=\tau+1}^{T-1} \beta_{jt} u_{it}, \quad \tau = \overline{1, T-2},$$

$$r_{ijT-1}(u) = \sum_{t=1}^{T-1} \beta_{jt} R_{jt}.$$

The solution of (4.107) is given, of course, by the formulae

$$z_{ij}(u) = \min_{\tau = \overline{0, T-1}} r_{ij\tau}(u), \quad (4.108)$$

$$v_j(u) = \min_{i=\overline{1,m}} \left[ c_{ij} + \frac{1}{b_j} z_{ij}(u) + u_{iT} \right]. \tag{4.109}$$

Thus problem $(4.100)-(4.105)$ reduces to the problem of maximizing the piecewise linear, concave function $\tilde{F}(u)$:

$$\max \left[ \tilde{F}(u) = \sum_{j+1}^{n} b_j v_j(u) - \sum_{i,t=1}^{m,T} \alpha_{it} u_{it} \right], \tag{4.110}$$

subject to the constraints

$$u_{it} \geqq l_{it}, \quad i = \overline{1,m}, \quad t = \overline{1, T-1}. \tag{4.111}$$

Let us consider the unconstrained maximization problem

$$\max_{u} F(u), \tag{4.112}$$

where the function

$$F(u) = \sum_{j=1}^{n} b_j v_j(u) - \sum_{i,t=1}^{m,T-1} {}' \alpha_{it} \max \{u_{it} - l_{it}\} - \sum_{i=1}^{m} \alpha_{iT} u_{iT}$$

is also piecewise linear and concave, and $v_j(u)$ is, as above, defined by formulae $(4.108)-(4.109)$.

Let $\tilde{u}^*$ denote a solution to problem $(4.112)$. Then it is easy to see that the solution $u^*$ to problem $(4.110)-(4.111)$ is given by the formulae

$$u_{it}^* = \max \{\tilde{u}_{it}^*, - l_{it}\}, \quad t = \overline{1, T-1},$$

$$u_{iT}^* = \tilde{u}_{iT}^*.$$

Thus the problem of determining the assignment of the consumers to the producers reduces to the unconstrained problem $(4.112)$ of maximizing a piecewise linear concave function, which can be solved by one of the versions of the optimization method with space dilation, namely the $r$-algorithm, successfully tested on problems of minimizing piecewise smooth functions [91, 23].

The problem of determining the supplies $x_{ijt}$ reduces to the problem of finding a solution to the primal problem $(4.92)-(4.99)$ on the basis of the obtained solution to the dual problem $(4.100)-(4.105)$.

We shall now describe a general method for finding a solution to the primal problem, and then show how to apply this method to our problem.

Suppose we are given a solvable convex programming problem of the form

$$\min_{x \in \Omega} f(x) \tag{4.113}$$

$$g_i(x) \leqq 0, \quad i = \overline{1,m}.$$

We suppose that the above problem satisfies the conditions of the Kuhn-Tucker theorem. Let us write the dual problem

$$\max_{\lambda \geq 0} \min_{x \in \Omega} L(x, \lambda), \tag{4.114}$$

where $L(x, \lambda) = f(x) + \sum_{i=1}^{n} \lambda_i g_i(x)$ is the Lagrange function.

It is easy to show that the function $\psi(\lambda) = \min_{x \in \Omega} L(x, \lambda)$ is concave. We shall assume that $\psi$ is proper, i.e. defined for all $\lambda \geq 0$. To this end it suffices to require that $\Omega$ be bounded.

Let $x(\lambda)$ denote a solution (in general, nonunique) to the problem $\min_{x \in \Omega} L(x, \lambda)$, i.e. $\psi(\lambda) = L(x(\lambda), \lambda)$. Then it is easy to show that $g(x(\lambda)) = (g_1(x(\lambda)), \ldots, g_m(x(\lambda)))$ is a subgradient of $\psi$ at $\lambda$.

Let $\lambda^*$ denote a solution to the dual problem and $x^*$ denote the solution to the primal problem corresponding to $\lambda^*$, i.e. the relations $\lambda_i^* g_i(x^*) = 0$, $i = \overline{1, m}$, are satisfied. It follows from the duality theorem that $\psi(\lambda^*) = f(x^*)$. Define the set $M(\lambda^*)$:

$$M(\lambda^*) = \{x(\lambda^*) \mid L(x(\lambda^*), \lambda^*) = f(x^*)\}. \tag{4.115}$$

Clearly, $M(\lambda^*)$ is a closed, convex set, and $x^* \in M(\lambda^*)$, since $L(x^*, \lambda^*) = f(x^*)$. Therefore, if $M(\lambda^*)$ consists of one element (for instance, if $L(x, \lambda^*)$ is strictly convex with respect to $x$), then to determine $x^*$ it suffices to solve the problem $\min_{x \in \Omega} L(x, \lambda^*)$.

We shall now consider the case when (4.113) is a linear programming problem. We shall assume that all $\lambda_i^* > 0$ (the constraints corresponding to $\lambda_i^* = 0$ are insignificant and can be deleted from the formulation of the primal problem). To solve the dual problem $\max_{\lambda} \psi(\lambda)$ we shall use the sub-gradient method

$$\lambda^{(k+1)} = \lambda^{(k)} + h_k \psi'(\lambda^{(k)}), \quad k \geq 1, \tag{4.116}$$

under the conditions

$$h_k \geq 0, \quad h_k \to 0, \quad \sum_{k=1}^{\infty} h_k = \infty, \tag{4.117}$$

which ensure the convergence of $\{\lambda^{(k)}\}$ to the solution $\lambda^*$. Define the sequence $\{x_N\}$ by the formula

$$x_N = \sum_{k=1}^{N} \mu_k^N x^{(k)}, \tag{4.118}$$

where $x^{(k)} = x(\lambda^{(k)})$ and $\mu_k^N = h_k / \sum_{i=1}^{N} h_i$. Since $\sum_{k=1}^{N} \mu_k^N = 1$, $\mu_k^N \geq 0$ and $\Omega$ is a convex set, we have $x_N \in \Omega$. We shall show that each accumulation point $\tilde{x}$ of the sequence $\{x_N\}$ is a solution to the primal problem (4.113).

If follows from the closedness of $\Omega$ that $\tilde{x} \in \Omega$. We shall demonstrate that $g(\tilde{x}) = 0$.

By the linearity of $g_i$, we have

$$g(x_N) = \sum_{k=1}^{N} \mu_k^N g(x^{(k)}) = \sum_{k=1}^{N} \mu_k^N \psi'(\lambda^{(k)}).$$

From (4.116) we get the equality

$$\sum_{k=1}^{N} \mu_k^N \psi'(\lambda^{(k)}) = \frac{\lambda^{(N+1)} - \lambda^{(I)}}{\sum_{i=1}^{N} h_i},$$

therefore, by taking account of the fact that $\lambda^{(k)} \to \lambda^*$ and $\sum_{i=1}^{N} h_i \to \infty$, we obtain $\lim_{n \to \infty} g(x_N) = 0$, from which the desired results follows.

We shall prove that $f(\tilde{x}) = f(x^*)$. It is easy to observe that for a linear programming problem the function $\psi$ is piecewise linear and convex, which yields the existence of a neighborhood $D(\lambda^*)$ of $\lambda^*$ such that $G(\lambda) \subset G(\lambda^*)$ for all $\lambda \in D(\lambda^*)$, where $G(\lambda)$ is the generalized gradient set of $\psi$ at $\lambda$. Since $\lambda^{(k)} \to \lambda^*$, $\lambda^{(k)} \in D(\lambda^*)$ for all sufficiently large $k$. Therefore with no loss of generality we may assume that all the sequence $\{\lambda^{(k)}\} \subset D(\lambda^*)$.

It was shown above that $\psi'(\lambda) = g(x(\lambda))$. It turns out that a converse, in a sense, assertion is true: if $\psi'(\lambda_0) = g(x(\lambda_1))$ then $x(\lambda_1) \in M(\lambda_0)$, where $M(\lambda_0)$ is defined analogously to $M(\lambda^*)$ in (4.115). It follows that, since $g(x^{(k)}) = \psi'(\lambda^{(k)}) \subset G(\lambda^*)$ we have $x^{(k)} \in M(\lambda^*)$. But $M(\lambda^*)$ is a convex set, therefore $x_N \in M(\lambda^*)$, and from the closedness of $M(\lambda^*)$ we obtain $\tilde{x} \in M(\lambda^*)$. Consequently, $f(\tilde{x}) = f(x^*)$, $x \in \Omega$ and $g(\tilde{x}) = 0$, which means that $x$ is a solution to problem (4.113), as required.

We shall now describe an application of the above method for finding a solution of the primal problem to the pair of problems (4.92)–(4.99) and (4.100)–(4.105). We start by selecting the set of consumers for which $x_{ijt}$ is found directly from the solution $v^*$, $u^*$, $z^*$ to problem (4.100)–(4.105).

Let $I^*(j)$ denote the set of indices $i$ for which the minimum in formula (4.109) is attained at $u = u^*$. By $T_i^*(j)$ we denote the set of indices $\tau$ for which the minimum in (4.108) is attained at $u = u^*$, $I_1$ denotes the set of consumers for which the set $I^*(j)$ consists of one element $i^*(j)$, and let $W$ be the set of the pairs of indices $(i, j)$, $i \in I^*(j)$, for which $T_i^*(j)$ consists of one element $\tau_{ij}^*$.

Let us define the set $I_2$ by the formula

$$I_2 = \{j \mid j \in I_1, (i(j), j) \in W\}.$$

It is not difficult to see that for $j \in I_2$

$$x_{ijt} = \begin{cases} 0 & \text{if } i \neq i^*(j), \\ 0 & \text{if } t \leq \tau_{ij}^*, \\ \beta_{it} & \text{if } t > \tau_{ij}^* \text{ and } i = i^*(j). \end{cases}$$

Thus the plan of supplies $\{x_{ijt}\}$ to the consumers in $I_2$ is determined directly from the solution to the dual problem (4.100)–(4.105). For determining the plan of supplies to the remaining consumers it is necessary to solve problem (4.92)–(4.99), in which one eliminates a priori the variables corresponding to the consumers $j$ from $I_2$, i.e. the variables $a_i$ and $\alpha_{it}$ are substituted by $(a_i - \sum_{j \in I_2} x_{ijT})$ and $(\alpha_{it} - \sum_{j \in I_2} x_{ijt})$, respectively, and the index $j$ ranges in $\tilde{I} = [\,1, n\,] \backslash I_2$.

It is easy to obtain the following estimate on $\tilde{n} = |\tilde{I}|$: $\tilde{n} \leq m\,T$. Therefore, having found the optimal assignment of the consumers, we arrive at a problem of the form (4.92)–(4.99) with a relatively small size $(n = \tilde{n})$, to the solution of which one can apply the method described above.

The Lagrange function of problem (4.92)–(4.99) will be constructed with respect to constraints (4.94) and (4.95), with the corresponding multipliers denoted by $u_{iT}$ and $u_{it}$, as above. To make the set $\Omega$, defined by the remaining constraints, bounded, we shall introduce an additional constraint $w_{it} \leq \alpha_{it}$, which is a consequence of all the constraints of problem (4.92)–(4.99). Then the dual problem is to find

$$\max_u \Psi(u), \tag{4.119}$$

where

$$\Psi(u) = \min_{x \in \Omega} L(x, u)$$

(for brevity, $x$ denotes here the set of the variables $\{x_{ijT}\}$, $\{y_{ij}\}$ and $\{w_{it}\}$), the set $\Omega$ is defined by the constraints

$$\sum_{i=1}^{m} x_{ijT} = b_j, \quad -\frac{1}{b_j} x_{ijT} + \sum_{\tau=0}^{T-1} y_{ij\tau} = 0;$$

$$w_{it} \leq \alpha_{it},$$

$$x_{ijT} \geq 0, \quad y_{ij\tau} \geq 0, \quad w_{it} \geq 0,$$

and the Lagrange function $L(x, u)$ is defined in a standard way.

It is easy to see that the solution $x(u)$ to the problem $\min\limits_{x \in \Omega} L(x, u)$ is determined as follows:

$$w_{ij} = \begin{cases} \alpha_{it} & \text{if } u_{it} + l_{it} \leq 0 \\ 0 & \text{if } u_{it} + l_{it} > 0, \end{cases}$$

$$y_{ij} = \begin{cases} 1 & \text{if } i = i^*(j) \text{ and } \tau = \tau^*(j), \\ 0 & \text{if } i \neq i^*(j) \text{ or } \tau \neq \tau^*(j), \end{cases}$$

where $i^*(j)$ and $\tau^*(j)$ are an arbitrary pair of indices $(i, \tau)$ for which the minimum is attained in the following expression

$$\min_{i=\overline{1,m}} [c_{ij} + u_{iT} + \min_\tau r_{ij\tau}(u)],$$

$r_{ij\tau}(u)$ is defined as in (4.108), and

$$x_{ijT} = \begin{cases} b_j & \text{if } i = i^*(j), \\ 0 & \text{if } i \neq i^*(j). \end{cases}$$

Thus the evaluation of the function $\Psi$ and its generalized gradient creates no difficulty. Consequently, for solving problem (4.119) one can use the subgradient method and obtain a solution to the original problem (4.92) − (4.99) by formula (4.118). The corresponding programs were worked out by N. G. Zhurbenko and L. V. Belaeva.

## 2. The Distribution of Coking Coals among Coking Plants

We solved this medium size problem for various initial data with a controlled accuracy. As it is well known, the efficiency of subgradient methods in comparison with finite methods of linear programming increases, as a rule, with an increasing size of the problem. Yet, even for that relatively small problem the subgradient methods with space dilation turned out to be rather efficient. The corresponding programs were elaborated and tested in practice by V. I. Gershovich [35].

The economic-mathematical model of coal distribution for coking, proposed by workers of the Informational and Computational Center of Minchermet USSR, is of the following form

minimize

$$\sum_j \sum_k \sum_i c_{ij} x_{ijk}, \tag{4.120}$$

subject to

$$\sum_i x_{ijk} \leqq Q_{jk}, \quad j \in J, \quad k \in K, \tag{4.121}$$

$$\sum_j x_{ijk} = \alpha_i^k \Theta_i, \quad i \in I, \quad k \in K, \tag{4.122}$$

$$\underline{z}_i \Theta_i \leqq \sum_j \sum_k z_{jk} x_{ijk} \leqq \bar{z}_i \Theta_i, \quad i \in I, \tag{4.123}$$

$$\underline{S}_i \Theta_i \leqq \sum_j \sum_k s_{jk} x_{ijk} \leqq \bar{S}_i \Theta_i, \quad i \in I, \tag{4.124}$$

$$P_i \Theta_i \leqq \sum_j \sum_k p_{ik} x_{ijk}, \quad i \in I, \tag{4.125}$$

$$x_{ijk} \geqq 0, \quad i \in I, \quad j \in J, \quad k \in K. \tag{4.126}$$

Here $i \in I = \{1, 2, \ldots, NI\}$ is an index of a coking chemical plant (CCP), $j \in J = \{1, 2, \ldots, N\}$ is an index of a coal-mine, and $k \in K = \{1, 2, \ldots, NK\}$ is an index of a type of coal. Thus $x_{ijk}$ is the volume of the $k$-th coal transported from the $j$-th colliery to the $i$-th CCP, $Q_{jk}$ is the production of coal $k$ by the $j$-th coal-mine, $\Theta_i$ is the total (with respect to the types) demand for coal of the $i$-th CCP, $\alpha_i^k$ is the quota of the $k$-th type of coal in the demand of the $i$-th CCP, the product $\alpha_i^k \Theta_i$ is the demand of the $i$-th CCP for the $k$-th type of coal, $z_{jk}(s_{jk})$ is the percentage of ash (sulphur) in the $k$-th type of coal mined at the $j$-th colliery, $p_{jk}$ is an analogous characteristic of plasticity, $\underline{Z}_i(\underline{S}_i)$ is the minimum admissible percentage of ash (sulphur) in the coal supplied to the $i$-th CCP, $\bar{Z}_i(\bar{S}_i)$ is the maximum admissible percentage of ash (sulphur) in the coal supplied to the $i$-th CCP, and $\underline{P}_i$ is an analogous characteristic of plasticity.

Let us rewrite problem $(4.120)-(4.126)$ is a more compact form, introducing obvious changes in the notation:

minimize

$$\sum_i \sum_j c_{ij} x_{ij}, \tag{4.127}$$

subject to

$$\sum_{i=1}^{N+1} x_{ij} = Q_j, \quad j \in J, \tag{4.128}$$

$$\sum_{j \in J_k} x_{ij} = \alpha_i^{(k)} \Theta_i, \quad i \in I, \quad k \in K, \tag{4.129}$$

$$\sum_j a_j^{(l)} x_{ij} \leqq b_i^l, \quad l \in L_1, \quad i \in I, \tag{4.130}$$

$$x_{ij} \geqq 0.$$

We note two aspects that have been used in passing to the above formulation.

1. Each coal-mine produces only one type of coal. Accordingly, the set $J$ of coal-mines decomposes into disjoint subsets, $J = \bigcup_k J_k$. (We observe that this limitation is not essential, since the problem can always be converted to the stipulated form by replacing a coal-mine that produces many types of coal with a corresponding group of collieries.)

2. From inequality (4.121) to equality (4.128) we pass by introducing formally in additional plant (with the index $i = NI + 1$). To this plant, in contrast with other CCP-s, do not correspond any technological constraints, and the cost of the coal transportation to this plant from any coal-mine is set equal to zero.

Moreover, observing that constraints (4.129) and (4.130) have an analogous structure and choosing in an appropriate way the coefficients $a_j^l$ (in fact we enlarge the number of rows of the matrix $\{a_j^l\} = A$, where $l$ indexes the rows and $j$ indexes the columns), we shall write the problem in the form

minimize

$$\sum_{i=1}^{NI} \sum_{j=1}^{NJ} c_{ij} x_{ij}, \tag{4.132}$$

subject to

$$\sum_{i=1}^{NI+1} x_{ij} = Q_j, \quad j \in J, \tag{4.133}$$

$$\sum_{j=1}^{NJ} a_j^l x_{ij} \leqq b_i^l, \quad i \in I, \quad l \in L_1, \tag{4.134}$$

$$\sum_{j=1}^{NJ} a_j^l x_{ij} = b_i^l, \quad i \in I, \quad l \in L_2, \tag{4.135}$$

$$x_{ij} \geqq 0, \tag{4.136}$$

where $L = L_1 \cup L_2$, $L_1 = \{1, \ldots, L\}$, $L_2 = \{L + 1, \ldots, NL\}$.

Consider the problem dual to $(4.132)-(4.136)$:
maximize

$$\left[\sum_{j=1}^{NJ} Q_j v_j - \sum_{i=1}^{NI} \sum_{l=1}^{NL} b_i^l u_{il}\right], \tag{4.137}$$

subject to

$$v_j - \sum_{l=1}^{NL} a_j^l u_{il} \leqq c_{ij}, \quad i \in I, \quad j \in J, \tag{4.138}$$

$$u_{il} \geqq 0, \quad i \in I, \quad l \in L_1. \tag{4.139}$$

To constraints (4.133) correspond the dual variables $v_j$, and with constraints $(4.134)-(4.135)$ we associate the dual variables $u_{il}$.

Let the values of the dual variables $\{\bar{u}_{il}, i \in I, l \in L_1\}$ be fixed. The resulting reduced problem (involving only the variables $v_j$) is of the form maximize

$$\left[\sum_{j=1}^{NJ} Q_j v_j - \sum_{i=1}^{NI} \sum_{l=1}^{NL} b_i^l \bar{u}_{il}\right],$$

subject to

$$v_j \leqq c_{ij} + \sum_{l=1}^{NL} a_j^l \bar{u}_{il}, \quad i \in I, \quad j \in J,$$

and has the trivial solution

$$v_j = \min_i \left\{c_{ij} + \sum_{l=1}^{NL} a_j^l \bar{u}_{il}\right\}. \tag{4.140}$$

In particular, if the fixed values $\bar{u}_{il}$ are such that $\bar{u}_{il} = u_{il}^*$, where $\{v_j^*, u_{il}^*\}$ is one of the solutions to problem $(4.137)-(4.139)$, then $v_j^*$ is determined by $u_{il}^*$ according to formula (4.140).

For finding optimal $\{v_j^*, u_{il}^*\}$ we shall use directly the developments and results of Sect. 4.1. We divide the constraints of problem $(4.132)-(4.136)$ into two groups. The first group consists of the constraints of the form (4.133), while the second one comprises the constraints of the form (4.134) and (4.135). The constraints of the second group, i.e. the technological constraints, will be used for constructing the Lagrange function

$$L(x, u) = \sum_{i,j} c_{ij} x_{ij} + \sum_{i,l} u_{il} \left(\sum_{i=1}^{NJ} a_j^l x_{ij} - b_i^l\right).$$

The optimal point $u^* = \{u_{il}^*\}$ of problem $(4.137)-(4.139)$ is a point of the maximum of the concave function

$$L^*(u) = \min_{x \in D} \left[ \sum_i \sum_j c_{ij} x_{ij} + \sum_i \sum_l u_{il} \left( \sum_j a_j^l x_{ij} - b_i^l \right) \right],  \tag{4.141}$$

subject to simple constraints $u_{il} \geqq 0$, $i \in I$, $l \in L_1$. Here the set $D$ is defined by constraints (4.135). From the expression

$$L^*(u) = \sum_i \sum_j c_{ij} x_{ij}^*(u) + \sum_i \sum_l u_{il} \left( \sum_j a_j^l x_{ij}^*(u) - b_i^l \right),$$

where $x^*(u) = \{x_{ij}^*(u)\}$ is a solution to problem (4.141), it follows that a subgradient of the function $L^*$ at a point $u$ is given by

$$g(u) = \left\{ \sum_j a_j^i x_{ij}^*(u) - b_i^l \right\}.$$

Having shown the way of calculating the subgradient $g(u)$ at any point $u$, we may use for maximizing $L$ the subgradient algorithm or any of its accelerated versions. The problem in question was solved by the subgradient algorithm with space dilation in the direction of the difference of two successive gradients [95].

We also note that it is a trivial task to find the solution $x^*(u)$ to problem (4.141), which is necessary for computing the subgradient $g(u)$. Indeed, let us reformulate problem (4.141) as

$$\min_{x \in D} \left[ \sum_{i=1}^{NI} \sum_{j=1}^{NJ} \left( c_{ij} + \sum_{l=1}^{NL} a_j^l u_{il} \right) x_{ij} - \sum_{i=1}^{NI} \sum_{l=1}^{NL} b_i^l u_{il} \right],  \tag{4.142}$$

where the feasible set $D$ is given by constraints (4.133), i.e.

$$\sum_{i=1}^{NI+1} x_{ij} = Q_j.  \tag{4.143}$$

This problem has a solution

$$x_{ij}^* = \begin{cases} Q_j & \text{for one } i \in I_j^*, \\ 0 & \text{otherwise,} \end{cases}$$

where

$$I_j^* = \{ i^* \mid \tilde{c}_{i^*j} = \min_i \tilde{c}_{ij} \},$$

$$\tilde{c}_{ij} = c_{ij} + \sum_{l=1}^{NL} a_j^l u_{jl}.$$

*The Case of an Inconsistent System of Constraints.* The computational experience has indicated that the constraints of the problem of distributing coking coals $(4.120)-(4.126)$ are frequently inconsistent. Such inconsistency arises from an excessive amount of ash (or sulphur) in the supplied coals in comparison with the admissible percentage. In this case we shall seek an approximate solution to the problem, i.e. a plan that optimizes the value of the objective function and satisfies constraints (4.121) and (4.122), but possibly violates constraints (4.123)–(4.125) by no more than some fixed quantity. It will be more convenient to use the formulation (4.127)–(4.131) of the problem and the dual problem (4.137)–(4.139).

The inconsistency of the original problem (4.127)–(4.131) with respect to constraint (4.130) is equivalent to the unboundedness of the dual problem resulting from the unboundedness of the feasible set of the dual problem (4.137)–(4.139) with respect to the variables $\{u_{il}, i \in I, l \in L_1\}$.

Let us consider the problem (called the quasi-dual problem in what follows):
maximize

$$\left( \sum_{j=1}^{NJ} Q_j v_j - \sum_{i=1}^{NI} \sum_{l=1}^{NL} b_i^l u_{il} \right), \tag{4.144}$$

subject to

$$v_j - \sum_{l=1}^{NL} Q_j^l u_{il} \leq c_{ij}, \quad i \in I, \quad j \in J, \tag{4.145}$$

$$u_{il} \geq 0, \quad i \in I, \quad l \in L_1, \tag{4.146}$$

$$u_{il} \leq d_{il}, \quad i \in I, \quad l \in L_1. \tag{4.147}$$

The quasi-dual problem differs from the dual problem (4.137)–(4.139) only by the additional constraint (4.147). In contrast with the dual, it has a finite optimal value.

Together with the quasi-dual we shall consider the problem (called the quasi-primal problem)
minimize

$$\sum_{i=1}^{NI} \sum_{j=1}^{NJ} c_{ij} x_{ij} + \sum_{i=1}^{NI} \sum_{l \in L_1} d_{il} y_i^l, \tag{4.148}$$

subject to

$$\sum_{i=1}^{NI} x_{ij} = Q_j, \quad i \in I, \tag{4.149}$$

$$\sum_{j=1}^{NJ} a_j^l x_{ij} - y_i^l \leq b_i^l, \quad i \in I, \quad l \in L_1, \tag{4.150}$$

$$\sum_{j=1}^{NJ} a_j^l x_{ij} = b_i^l, \quad i \in I, \quad l \in L_2, \tag{4.151}$$

$$x_{ij} \geq 0, \quad y_i^l \geq 0, \quad i \in I, \quad j \in J, \quad l \in L_1. \tag{4.152}$$

The quasi-primal (4.148)–(4.152) and the quasi-dual (4.144)–(4.147) constitute a pair of problems dual to each other, with the additional constraint (4.147) of problem (4.144)−(4.147) corresponding to the additional variables $y_i^l$ of problem (4.142)−(4.152).

Analogously to the preceding developments, we shall solve the quasi-dual problem (4.144)−(4.147) by reducing it to the problem of maximizing the concave function

$$L_R^* = \min_{x \in D}\,[L(x, u) - R(u)],$$

where

$$R(u) = \sum_{u=1}^{NI} \sum_{l \in L} r_{il}(u_{il} - d_{il})^+$$

is a penalty function taking account of constraints (4.147), which were introduced in the transition from the dual to the quasi-dual problem $(r_{il} > 0)$.

Similarly as above, we find a vector $u^*$ satisfying

$$L_R^*(u^*) = \max_{u_i^l \geqq 0} L_R^*(u) \tag{4.153}$$

by using the subgradient algorithm with space dilation. Then by formulae (4.140) we find the remaining component $v^*$ of the solution to the quasi-dual problem. In this way we obtain the vector $\{u^*, v^*\}$, which is a solution to problem (4.144)−(4.147) (cf. Sect. 4.1).

Let $\{x^*, y^*\}$ denote a solution to the quasi-primal problem (4.148)−(4.152). If the solutions to problems (4.144)−(4.147) and (4.153) coincide (and this is a true if the penalty parameters $r_{il}$ are properly chosen), then $y_i^l \leqq d_i^l$ [9]. But the variables $y_i^l$ in fact express a measure of the violation of constraints (4.140). This enables one to specify in some way the possible extent of the violation of these constraints by choosing the penalty coefficients $r_{il}$.

*The Solution of the Primal Problem.* Up till now the main attention has been devoted to finding a solution $\{u^*, v^*\}$ to problem (4.137)−(4.139), which is dual to the original problem (4.127)−(4.131). Since we are ultimately interested in an optimal solution $x^*$ to the primal problem, we shall now dwell upon the question of how to obtain it (or, equivalently, how to recover it) from the known vector $\{u^*, v^*\}$. We obtain the vector $x^*$ by solving the original problem (4.127)−(4.131) and taking account of the information about the solution $\{u^*, v^*\}$ to problem (4.137)−(4.139). This information enables one to significantly reduce the original problem. Indeed, undetermined are only those components $x_{ij}^*$ of the vector $x^*$ for which the corresponding constraints of (4.138) are satisfied as strict equalities if one substitutes the values $\{u^*, v^*\}$ in (4.138). The remaining components of the vector $x^*$ are known a priori to have zero values. Additionally, those constraints of the primal problem to which correspond the null components of the vector $\{u^*, v^*\}$ are inessential, i.e. they may be left out of consideration.

Actually, if one takes account of the limited accuracy of both the method and the computer calculations, expressions such as "satisfied as strict equalities" and "the null components of the vector $\{u^*, v^*\}$" should be understood in a somewhat different sense. Thus, the unknown components $x_{ij}^*$ are included in the reduced problem if the corresponding constraint in (4.138) of the dual problem is satisfied as equality with some accuracy $\varepsilon > 0$, i.e. if

$$0 \leqq c_{ij} - v_j^* = \sum_{l=1}^{NL} a_j^l u_{il}^* \leqq \varepsilon.$$

Similarly, a constraint of the primal problem is not included in the reduced problem if the corresponding component of the optimal dual vector is no larger than some small $\delta > 0$.

To sum up, the recovery of the primal variables consists only in finding the volumes of transports corresponding to the links established by solving the dual problem. The actual dimension of the reduced problem is equal to the number of such links.

As shown by our experimental computations, the number of the links obtained is smaller by an order than the number of variables in the original problem. Consequently, the reduced problem has a small dimension and can be easily solved by any finite method. To this end a simple variant of the simplex method was used. The time required for solving the reduced problem comprised $5-10\%$ of the overall expenditure of the computer time.

Finally, we remark that in the case of inconsistency of the original system of constraints all the results of this section remain valid if one replaces everywhere the primal problem $(4.127)-(4.131)$ by the quasi-primal problem $(4.148)-(4.152)$, the dual $(4.137)-(4.139)$ by the quasi-dual $(4.144)-(4.147)$, and the vector $x^*$ by the vector $(x^*, y^*)$.

*The Results of Experimental Computations.* As indicated above, problem (4.153), which is central in our approach, was solved by the subgradient algorithm with space dilation in the direction of the difference of two successive gradients [95]. To this end the version of the algorithm with a symmetric matrix, proposed in [96], was used, with some modifications connected with the choice of the moment of a consecutive space dilation and with the stepsize selection.

Let us briefly address these questions. The algorithm was allowed to take several steps along a search direction chosen in the current space. The movement along a given direction was terminated at a point at which it was known that a further step along this direction would have led to a worsening in the function value (at such a point the projection of the gradient on the search direction is negative). At this moment occurred a consecutive space dilation along the difference of the latest gradient and the gradient which determined the current search direction.

Clearly, such a strategy may be interpreted as an approximate directional maximization. We did not strive to attain monotonicity with respect to the function values, since the requirement of monotonicity led to a significant increase in the number of function evaluations and did not essentially influence the speed of convergence.

The stepsize values were controlled by the number of steps taken along the chosen direction. After every $k_1$ steps along this direction before the moment of a consecutive space dilation, the stepsize coefficient was increased by a factor of $q_1 \geqq 1$. If only one step was taken in the chosen direction, then the stepsize coefficient was decreased by a factor of $q_2 \leqq 1$.

In the computations, the following two sets of parameter values were used: (1) $k_1 = 5$, $q_1 = 1.2$, $q_2 = 0.95$, (2) $k_1 = 3$, $q_1 = 1.2$, $q_2 = 1$.

In the second case the decrease of shifts in the original space took place only due to the decrease of the matrix of space transformation. The computations were carried out for initial data provided by Minchermet USSR and Ukrglavugl. The algorithm was implemented as a program in FORTRAN-IV of DOS ES.

The dimension of the problem was given by the number of CCP-s $NI = 9$, the number of coal-mines $NJ = 72$, and the number of "technological" constraints on CCP $NL = 8$. Thus the original problem had $NI \times NJ = 648$ variables, while the gradient descent took place with respect to $N = NJ \times NL = 72$ dual variables $u_{il}$. As shown by the experimental calculations, the computation of $u^*$ yielding correct links in the original problem required, for various initial data, no more than $500 - 600$ iterations (i.e. about $8N$). This involved $700 - 1200$ evaluations of the function maximized, i.e. on the average no more than two steps per line search. The value of the coefficient of space dilation $\alpha = 3$ turned out to be preferable to the value $\alpha = 2$.

We give in Table 1 an illustration of the change of the function value in one experiment with the initial point $u = 0$, the initial stepsize $h = 50$, the coefficient of dilation $\alpha = 3$, $k_1 = 3$, $q_1 = 1.2$ and $q_2 = 1$.

**Table 1**

| $N_{\text{iter}}$ | $f \cdot 10^{-6}$ | $N_{\text{iter}}$ | $f \cdot 10^{-6}$ | $N_{\text{iter}}$ | $f \cdot 10^{-6}$ |
|---|---|---|---|---|---|
| 0 | −0.171 | 200 | 0.386 | 400 | 0.41211 |
| 20 | 0.216 | 220 | 0.406 | 420 | 0.41336 |
| 40 | 0.309 | 240 | 0.413 | 440 | 0.41476 |
| 60 | 0.334 | 260 | 0.4146 | 460 | 0.41523 |
| 80 | 0.349 | 280 | 0.4151 | 480 | 0.41555 |
| 100 | 0.359 | 300 | 0.4152 | 500 | 0.41566 |
| 120 | 0.365 | 320 | 0.4155 | 520 | 0.41575 |
| 140 | 0.369 | 340 | 0.41561 | 540 | 0.41585 |
| 160 | 0.374 | 360 | 0.41569 | 570 | 0.41588 |
| 180 | 0.378 | 380 | 0.41564 | 600 | 0.41592 |

In total 693 function evaluations were required. The time of computation on an M-4030 computer was equal to 55 min.

As shown by the experiments, the time spent on matrix updatings was three times longer than the time taken up by function evaluations. Therefore, in a natural way arises the problem of storing the information about space transformations in the form of a set of vectors and scalar products, instead of the usual matrix form (which appears redundant). Even if one uses the matrix form, then one should organize in a most efficient way (with respect to the number of operations) the block of matrix updatings at successive space dilations.

Finally, we point out the experimentally observed worsening of convergence at some stage of computations. In the example given above this occured between the 360-th and 400-th iterations. This worsening of the convergence properties of the algorithm is connected with the loss of positive definiteness of the matrix of transformations due to the limited precision of computer calculations, which starts to have a significant effect when the determinant of the matrix becomes small. In this way the information about the space transformation deteriorates and one has to reset the matrix by setting it equal to the unit matrix. At this moment the convergence with respect to function values worsens. To mitigate this effect, the diagonal elements of the matrix were slightly corrected so as to improve its positive definiteness. Seemingly, an effective means is double precision of calculations.

## 3. The Problem of Allocating Passenger Aircrafts to Airlines [105]

A simplified mathematical model of the problem (the so-called static model), formulated for a given year, is of the following form:
minimize the total yearly expenditure

$$c = \sum_{j \in J} \sum_{i \in I} c_{ij} x_{ij}, \tag{4.154}$$

subject to the constraints

$$\sum_{i \in I} p_{ij} x_{ij} \geqq d_j, \quad j \in J, \tag{4.155}$$

$$\sum_{i \in I} v_{ij} x_{ij} \geqq b_j, \quad j \in J_1, \tag{4.156}$$

$$\sum_{j \in J_k} \sum_{i \in I} v_{ij} x_{ij} \leqq B_k, \quad k \in K, \tag{4.157}$$

$$\sum_{j \in J} x_{ij} \leqq a_i, \quad i \in I_1, \tag{4.158}$$

$$\sum_{j \in J} x_{ij} \geqq r_i, \quad i \in I_2, \tag{4.159}$$

$$x_{ij} \geqq 0, \quad i \in I, \quad j \in J, \tag{4.160}$$

where $x_{ij}$ is the number of aircrafts of the $i$-th type allocated to line $j$, $c_{ij}$ is the total yearly outlay on an aircraft of type $i$ on the $j$-th line (in rubles per year), $p_{ij}$ is the mean number of passengers carried yearly by an aircraft of type $i$ on the $j$-th line (in passengers per year), $d_j$ is the minimum volume of transport on line $j$, $v_{ij}$ is the number of flights of an aircraft of the $i$-th type on the $j$-th line (in flights per year), $b_j$ is the minimum number of flights on line $j \in J_1 \subset J$, $J_k$ is the set of lines using airport $k$, $B_k$ is the total service capacity of the $k$-th airport, $a_i$ is the number of available aircrafts of type $i$, and $r_i$ is the minimum number of aircrafts of the $i$-th type that must be allocated.

Problem $(4.154)-(4.160)$ is a large scale linear programming problem (the number of variables equals approximately $13 \cdot 1400$, and the number of constraints exceeds $2 \cdot 1400$). A most suitable approach to the solution of such problems consists in using decomposition techniques together with finite or iterative methods for solving the separate blocks [77].

Let us write the dual of $(4.154)-(4.160)$

$$\max_{Z,W,U \geqq 0} \psi(Z,W,U), \tag{4.161}$$

where

$$\psi(Z,W,U) = \min_{x \in R} \left\{ \sum_{j \in J} \sum_{i \in I} c_{ij} x_{ij} \right.$$
$$+ \sum_{k \in K} Z_k \left( \sum_{j \in J_k} \sum_{i \in I} v_{ij} x_{ij} - B_k \right)$$
$$+ \sum_{i \in I_1} W_i \left( \sum_{j \in J} x_{ij} - a_i \right)$$
$$\left. + \sum_{i \in I_2} U_i \left( r_i - \sum_{j \in J} x_{ij} \right) \right\}.$$

The set $R$ is defined by the system of inequalities (4.155), (4.156). It is easy to observe that

$$\psi(Z,W,U) = \varphi(Z,W,U) - \sum_{k \in K} B_k Z_k - \sum_{i \in I_1} a_i W_i + \sum_{i \in I_2} r_i U_i,$$

where

$$\varphi(Z,W,U) = \min_{x \in R} \sum_{j \in J} \sum_{i \in I} \tilde{c}_{ij}(Z,W,U) x_{ij},$$

$$\tilde{c}_{ij}(Z,W,U) = c_{ij} + \sum_{k \in K} v_{ij} Z_k \delta_{jk} + \tilde{W}_i - \tilde{U}_i,$$

$$\delta_{jk} = \begin{cases} 1 & \text{if } j \in J_k, \\ 0 & \text{if } j \notin J_k, \end{cases}$$

$$\tilde{W}_i = \begin{cases} W_i & \text{if } i \in I_1, \\ 0 & \text{if } i \notin I_1, \end{cases}$$

$$\tilde{U}_i = \begin{cases} U_i & \text{if } i \in I_2, \\ 0 & \text{if } i \notin I_2. \end{cases}$$

Let us denote by $R_j$ the set defined by inequalities (4.155) and (4.156) for a fixed index $j$. We obtain

$$\varphi(Z, W, U) = \sum_{j \in J} \min_{x \in R_j} \sum_{i \in I} \tilde{c}_{ij}(Z, W, U)\, x_{ij},$$

i.e.

$$\varphi(Z, W, U) = \sum_j \sum_i \tilde{c}_{ij}(Z, W, U)\, y_{ij}^*,$$

where $y_{ij}^*$ is a solution to the following problem

$$\min_{y_{ij} \in R_j} \sum_{i \in I} \tilde{c}_{ij} y_{ij}, \quad j \in J. \tag{4.162}$$

For $j \in J \backslash J_1$ a solution to problem (4.162) is given by the formula

$$y_{ij}^* = \begin{cases} 0 & \text{if } i \neq i^*, \\ d_i/p_{ij} & \text{if } i = i^*, \end{cases}$$

where $i^*$ is an index for which the quotient $\tilde{c}_{ij}/p_{ij}$ attains its minimum value.

For $j \in J_1$ problems (4.162) constitute linear programming problems of small dimension ($10-13$ variables and 2 constraints), and it is advisable to solve them by one of the modifications of the simplex method.

To reduce problem (4.162) to an unconstrained maximization problem, it suffices to make the transformation of variables

$$Z_k = |Z_k'|, \quad W_i = |W_i'|, \quad U_i = |U_i'|.$$

Although under such a transformation the function $\tilde{\psi}(Z_k', W_i', U_i')$ $= \psi(|Z_k'|, |W_i'|, |U_i'|)$ is nonconcave, still all its local maxima correspond to the solution of problem (4.161).

The method for solving problem $(4.154)-(4.160)$ consists of two stages. On the first stage the solution $\{Z^*, W^*, U^*\}$ to problem (4.161) is found by means of the $r$-algorithm. The generalized gradient of the function $\psi(Z, W, U)$, which is used in the $r$-algorithm, is calculated by the formulae

$$\psi_{Z_k}' = \sum_{j \in J_k} \sum_{i \in I} y_{ij}^* v_{ij} - B_k, \quad k \in K,$$

$$\psi_{W_i}' = \sum_{j \in J} y_{ij}^* - a_i, \quad i \in I_1,$$

$$\psi_{U_i}' = \sum_{j \in J} y_{ij}^* + r_j, \quad i \in I_2.$$

The analysis of the results of the first stage of the method determines those airlines for which $y_{ij}^*(Z^*, W^*, U^*) = x_{ij}$ is the solution to problem $(4.154)-(4.160)$, i.e. the lines for which the optimal allocation is found directly from the solution to the dual problem (4.161). Those airlines are eliminated from problem $(4.154)-(4.160)$. The resulting problem (its structure coincides with $(4.154)-(4.160)$) will be denoted by $\overline{(4.154)-(4.160)}$ and the corresponding dual by $\overline{(4.161)}$. It is easy to observe that the number of lines in problem $\overline{(4.154)-(4.160)}$ will be equal to the total number of constraints $(4.157)-(4.159)$. Thus the first stage will find an optimal solution for most airlines.

On the second stage problem $\overline{(4.161)}$ is solved by the method of generalized gradient descent. To obtain a solution to problem $\overline{(4.154)-}$ $\overline{(4.160)}$, i.e. to find an optimal allocation of aircrafts to the remaining lines, the following formula is used

$$x_{ij} = \frac{1}{N} \sum_{k=1}^{N} y_{ij}^*(Z^{(k)}, W^{(k)}, U^{(k)}),$$

where $Z^{(k)}$, $W^{(k)}$ and $U^{(k)}$ denote the values of the dual variables at the $k$-th iteration, and $N$ is the total number of iterations of the generalized gradient descent.

The above-described algorithm for solving problem $(4.154)-(4.160)$ has been implemented in ALGOL-60 on a BESM-6 computer. Since the implemented program is designed for solving a whole series of problems of the form $(4.154)-(4.160)$, it is provided with segments for preparing initial data, blocks for communicating with external storage devices, blocks for handling final results, etc.

To economize the memory of a computer it is advisable to store only the parameters of aircrafts, lines and airports. The constant $c_{ij}$, $p_{ij}$ and $v_{ij}$ can be comparatively easily computed during the run of the algorithm (the calculation of one constant requires about ten arithmetic operations). Owing to the exploitation of specific features of the problem, the algorithm described above can find its solution by using only the internal memory of BESM-6 (about 32,000 words).

The computations, which were performed by a program developed by N. G. Zhurbenko and T. B. Belykh, exhibited high effectiveness of the $r$-algorithm. The solution of the dual problem required about $80-100$ iterations for attaining the relative accuracy of $10^{-6}$ with respect to function values. For most airlines the solution was determined on the first stage, while on the second stage only $40-50$ lines were considered. The total computation time was $40-50$ min.

## 4. The Stochastic Transportation Problem [6]

The mathematical model of the problem consists in the following: minimize

$$\left[ \sum_{i,j=1}^{m,n} c_{ij} x_{ij} + \sum_{j=1}^{n} r_j E\left(\xi_j - \sum_{i=1}^{m} x_{ij}\right)^+ \right], \tag{4.163}$$

subject to

$$\sum_{j=1}^{n} x_{ij} \leq a_i, \quad i = \overline{1, m}, \tag{4.164}$$

$$x_{ij} \geq 0, \quad i = \overline{1, m}, \quad j = \overline{1, n}, \tag{4.165}$$

where $m$ is the number of suppliers (plants), $n$ is the number of consumers (building sites), $c_{ij}$ is the unit transportation cost from the $i$-th plant to the $j$-th building site, $a_i$ is the volume of production of the $i$-th plant, $\xi_j$ is a random variable describing the demand of the $j$-th building site, $r_j$ is the coefficient of penalty for not satisfying the demand of the $j$-th building site, $x_{ij}$ is the sought volume of transport from the $i$-th plant to the $j$-th building site, $t^+ = \max\{0, t\}$, and $E$ denotes the mathematical expectation.

To sum up, the problem consists in finding transport volumes that yield the minimum of the sum of the transportation costs and the expected losses resulting from deficits in supplies.

This problem is a special case of the two-stage stochastic programming problem [103]. The random variables $\xi_j$ are assumed to be independent and have probability density functions $p_j(z)$.

**Lemma 4.3.** *Problem* (4.163)$-$(4.165) *is a convex programming problem.*

*Proof.* For proving the assertion it suffices to establish the convexity of the function

$$f(t) = E(\xi - t)^+,$$

where $\xi$ is a random variable having the density $p(z)$. Clearly,

$$f(t) = \int_{-\infty}^{\infty} (z - t)^+ p(z)\, dz = \int_{t}^{+\infty} (z - t)\, p(z)\, dz,$$

hence $f'(t) = -\int_t^\infty p(z)\, dz$, $f''(t) = p(t)$. Since $p(t) \geq 0$, the last equality implies the convexity of $f$. $\square$

The solution of problem (4.163)$-$(4.165) will be based on applying a modification of the minimization method with space dilation (the $r$-algorithm) to the dual of (4.163)$-$(4.165).

Let us denote by $u$ the dual variables corresponding to constraints (4.164). Then the dual problem is of the following form:

$$\max_{u \geq 0} \psi(u), \tag{4.166}$$

where

$$\psi(u) = \min_{x \geq 0} L(x, u).$$

Here $x$ denotes the variables $\{x_{ij}\}$, and $L(x, u)$ is the Lagrange function

$$L(x, u) = \sum_{i,j=1}^{n,m} c_{ij} x_{ij} + \sum_{j=1}^{n} r_j E\left(\xi_j - \sum_{i=1}^{m} x_{ij}\right)^+ + \sum_{i=1}^{m} u_i \left(\sum_{i=1}^{m} x_{ij} - a_i\right).$$

It is easy to observe that

$$\psi(u) = \sum_{j=1}^{n} \varphi_j(u) - \sum_{i=1}^{m} a_i u_i,$$

where

$$\varphi_j(u) = \min_{x_{ij} \geq 0} \left[ \sum_{j=1}^{n} (c_{ij} + u_i) \, x_{ij} + r_j \, E \left( \xi_j - \sum_{i=1}^{m} x_{ij} \right)^+ \right]. \tag{4.167}$$

Let

$$v_j = \sum_{i=1}^{m} x_{ij}$$

and let $i(j)$ denote any index $i \in [1, m]$ for which the minimum is attained in the following expression:

$$\min_{i \in \overline{1,m}} (c_{ij} + u_i). \tag{4.168}$$

Then it is easy to show that the values of $x_{ij}$ which minimize the bracketed term in formula (4.167) are the following

$$x_{ij}(u) = \begin{cases} v_j(u) & \text{if } i = i(j), \\ 0 & \text{if } i \neq i(j). \end{cases} \tag{4.169}$$

Therefore

$$\varphi_j(u) = \tilde{c}_j \, v_j + r_j \, E \, (\xi_j - v_j)^+, \tag{4.170}$$

where

$$\tilde{c}_j = c_{i(j)j} + u_{i(j)}.$$

**Lemma 4.4.** *The variables $v_j(u)$ in formula (4.169) are determined in the following way: if $\tilde{c}_j \geq r_j$ then $v_j = 0$, otherwise $v_j$ is the solution to the equation*

$$r_j(1 - F_j(v)) = \tilde{c}_j, \tag{4.171}$$

*where $F_j(z)$ is the distribution function of the variable $\xi_j$.*

*Proof.* For brevity the index $j$ in formula (4.171) will be omitted. Consider the function

$$f(v) = \tilde{c} v + r \, E(\xi - v)^+ = \tilde{c} v + r \int_{v}^{\infty} (z - v) \, p(z) \, dz.$$

We get

$$f'(v) = \tilde{c} - r \int_{v}^{\infty} p(z) \, dz = c - r(1 - F(z)),$$

which yields the desired conclusion.  $\square$

Lemma 4.4 and relations (4.170) and (4.171) fully determine an algorithm for evaluating the function $\psi$. A generalized gradient $\psi'(u)$ of the function $\psi$ at $u$ is given by the formula $\psi'(u) = g(u)$, where

$$g(u) = (g_1(u), \ldots, g_m(u)),$$

$$g_i(u) = \sum_{j=1}^{n} x_{ij}(u) - a_i, \quad i = \overline{1, m}.$$

The dual problem (4.166) can be solved by any minimization method that does not require differentiability of the objective function (from the above definition of the function $\psi$ one easily concludes that $\psi$ is a concave piecewise linear function). One of such methods, having accelerated convergence, is the minimization method with space dilation in the direction of the difference of two successive gradients (the $r$-algorithm), which was validated in the solution of numerous practical problems.

To solve the dual problem (4.166) by the $r$-algorithm one should preliminary reduce it to an unconstrained optimization problem. This can be easily achieved by the transformation of variables $u = |u'|$.

Suppose that the $r$-algorithm has obtained a point $\tilde{u}$, approximating the solution $u^*$ to problem (4.166). We shall now consider the problem of finding the solution $x_{ij}$ to the original problem (4.163) − (4.165). It is clear from Lemma 4.4 that (in the case of strict monotonicity of the distribution functions $F_j$) the variables $v_j(u^*)$ are uniquely determined by the values of the dual variables $u^*$, hence one can show [6] that

$$v_j(u^*) = \sum_{i=1}^{m} x_{ij}^*,$$

i.e. $v_j(u^*)$ is the optimal value of the supplies delivered to the $j$-th consumer. Since we have assumed the existence of regular probability densities $p_j$, $v_j(u)$ is a continuous function of $u$, as can be easily deduced from Lemma 4.4. Therefore $v_j(\tilde{u})$ are estimates of $v_j(u^*)$, and can be used for reducing the nonlinear problem (4.163) − (4.165) to the classical transportation problem. For solving the latter problem one may use either the well-known algorithms or (which is more natural in the proposed "iterative" approach to the solution of the problem) the method of averaging (4.118) in the course of solving the corresponding dual by the subgradient algorithm.

The above-described mathematical model originated from a practical problem connected with supplying building materials to the building sites of Ukraine. In this problem we had $n \sim 100$ and $m = 30$. The random demands $\xi_j$ had Gaussian distributions, which enabled us to reduce the solution of Eq. (4.171) to the solution of equations of the form $\mathrm{erf}(z) = d_j$, where erf is the error function

$$\mathrm{erf}(z) = \frac{1}{\sqrt{\pi}} \int_0^z e^{-t^2} dt, \quad 0 \le d_j \le 1.$$

Therefore, for solving Eq. (4.171) it suffices to tabulate the inverse function $\mathrm{erf}^{-1}(y)$.

## 4.5 Application of $r$-Algorithms to Nonlinear Minimax Problems

We shall consider the problem of minimizing functions of the form $f(x) = \max\limits_{1 \le i \le n} \varphi_i(x)$, where $\varphi_i$ are smooth functions and $x \in E_n$. If $\varphi_i$ are convex for $i = 1, \ldots, m$, then as a subgradient $g_f(x)$ one may take a subgradient $g_{\varphi_{i*}}(x)$, where $i^*$ is an index of a function $\varphi_{i*}$ satisfying $f(x) = \varphi_{i*}(x)$.

In view of the importance of minimax problems much attention is presently given to the elaboration of minimax methods. Several such methods for certain classes of problems were proposed in the works of V. F. Demyanov, B. N. Pshenichny, E. G. Evtushenko and others (see [18]).

In theory, finding the minimax can be reduced to the solution of the following convex programming problem of minimizing $y$, subject to the constraints $\varphi_i(x) - y \le 0$, $i = 1, \ldots, m$.

For solving this problem one may use iterative methods of feasible directions [106]. However, this way of solving the problem is burdened with the following difficulties. Firstly, finding constrained minima by the methods of feasible directions requires solving at each iteration auxiliary linear or quadratic programming problems. Secondly, to ensure convergence of the methods of feasible directions one must take precautions against zig-zags (jamming). Thirdly, each iteration of the methods of feasible directions needs a directional minimization, which requires much effort, since one has to find the minimum of a nonsmooth function in a situation where each function evaluation requires considerable computational work. And finally, neither in theory nor in practice has the speed of convergence of such methods been sufficiently investigated.

Simpler from the computational point of view is the subgradient algorithm, which, in theory, can solve convex minimax problems [76]. Yet the speed of convergence of this method is not sufficiently high, especially for gully-type functions, which are characteristic of minimax problems. In this case the minimum is, as a rule, attained for equal values of several functions, with the number $m_1$ of such functions determining the dimension $n - n_1 + 1$ of the ridge.

The convergence of the subgradient algorithm can be accelerated by dilating the space in the direction of a generalized gradient or in the direction of the difference of two generalized gradients evaluated at two consecutive points [95]. The subgradient algorithm with space dilation determines a minimizing sequence by evaluating a generalized gradient of $f$ at each point of this sequence and applying the operator of space dilation.

When analyzing the efficiency of computational methods one usually assumes that an evaluation of the gradient of a function of $n$ variables requires as much time as $n + 1$ evaluations of the function. If we assume that each function $\varphi_i$ is continuously differentiable and that the work per its evaluation does not depend on the index $i$ and is $(n + 1)$-times smaller than the work par its gradient evaluations, then the complexity of evaluating $f$ constitutes a fraction of $m/(1+n)$ of the complexity of evaluating its subgradient. Thus with increasing $m$ the relative complexity of evaluating a subgradient $g_f$, in comparison with evaluating $f$, decreases. Therefore, in order to ensure high efficiency of algorithms for minimax problems, one should especially aim at simplifying procedures for directional minimization which require evaluating the max function. Guided by this observation, L. P. Shabashova and the author proposed and investigated experimentally several modifications of the $r$-algorithm, differing in stepsize selection rules and the accuracy of directional minimization [93].

Let us consider one such modification. The algorithm starts from an arbitrary point $x_0 \in E_n$, setting $\tilde{g}_0 = 0 \in E_n$ and $B_0 = I_n$, where $I_n$ is the $n \times n$ identity matrix. Suppose that the first $k$ iterations ($k = 1, 2, \ldots$) resulted in points $x^1, x^2, \ldots, x^k$, a vector $\tilde{g}_k$ and a matrix $B_k$. To determine the point $x^{k+1}$ we carry out the following operations.

1. We calculate the function value at the $k$-th point

$$f(x_k) = \max_{i \in \overline{1,m}} \varphi_i(x_k) = \varphi_{i*}(x_k).$$

2. We calculate the subgradient of $f$ at $x^k$

$$g_f(x_k) = g_{\varphi_{i*}}(x_k).$$

3. We find the subgradient $\bar{g}_k$ of the function $\psi_k$ at the point $y_k$ in the transformed space by the formula

$$\bar{g}_k = g_{\psi_k}(y_k) = B_k^* g_f(x_k),$$

where

$$\psi_k(y) = f(A_k^{-1} y) = f(B_k(y)),$$

$A_k$ is the operator of space transformation, $B_k$ is the inverse of $A_k$ and $B_k^*$ is the transpose of $B_k$.

4. We compute the difference of the two subgradients $r_k = \bar{g}_k - \tilde{g}_k$.
5. We calculate the ratio of the norms of $r_k$ and $g_k$

$$\beta_k = \frac{\|r_k\|}{\|g_k\|},$$

which characterizes the change of the direction of the subgradient at $x_{k-1}$ with respect to the direction of the subgradient at $x_k$ in the transformed space.

6. We compare $\beta_k$ with a fixed parameter $q_1$:
   a) if $\beta_k \leq q_1$ then we generate the next point

$$x_{k+1} = x_k - h_k B_k \frac{\tilde{g}_k}{\|\tilde{g}_k\|}$$

corresponding to the shift in the transformed space of point $y$ in the same direction and with the same stepsize $h_k$ as at the $k$-th iteration and we pass to the $(k + 2)$-nd iteration;

b) if $\beta_k > q_1$, i.e. if the subgradient changed significantly, then we execute the following steps.

7. We compute the vector $\xi_{k+1} = r_k / \| r_k \|$, which will determine the direction of space dilation.

8. We compute the operator $B_{k+1}$, the inverse of $A_{k+1}$, by the formula $B_{k+1} = B_k R_{1/\alpha}(\xi_{k+1})$.

9. We calculate the subgradient of the function $\psi_{k+1}$ at the point $\tilde{y}_{k+1}$ corresponding to $x_k$ under the $(k + 1)$-st space transformation

$$g_{k+1} = B_{k+1}^* g_f(x_k).$$

This vector determines the direction of movement in the transformed space.

10. We change the stepsize $h_{k+1} = h_k q_2$, where $0 < q_2 \leq 1$.

11. We calculate the next point

$$x_{k+1} = x_k - h_{k+1} B_{k+1} \frac{\tilde{g}_{k+1}}{\| \tilde{g}_{k+1} \|},$$

which corresponds to the point obtained by shifting in the transformed space form the point $y_k = A_k x_k$ in the direction $- \tilde{g}_{k+1}/\| \tilde{g}_{k+1} \|$ with the stepsize $h_{k+1}$. Then we pass on to the $(k + 2)$-nd iteration $(k = 1, 2, \ldots)$.

The efficiency of the above-described algorithm was investigated by implementing standard programs in the assembler language of the Minsk-22 computer and performing numerous calculations for minimax problems. For numerical experiments problems of the following form were used

$$\min_{x \in E_n} f(x),$$

where $f(x) = \max_{i \in \overline{1,m}} \varphi_i(x)$, $\varphi_i(x) = b_i \sum_{j=1}^{n} (x^{(j)} - a_{ij})^2$, $x = (x^{(1)}, \ldots, x^{(n)})$, $\{b_i\}$ is an $m$-vector and $\{a_{ij}\}$ is an $m \times n$ matrix.

As an illustration we report computational results for the following problem: $n = 5$, $m = 10$, $\{b_i\} = \{1, 5, 10, 2, 4, 3, 1.7, 2.5, 6, 4.5\}$, with the transpose of the matrix $\{a_{ij}\}$ given below

$$\begin{bmatrix} 0 & 2 & 1 & 1 & 3 & 0 & 1 & 1 & 0 & 1 \\ 0 & 1 & 2 & 4 & 2 & 2 & 1 & 0 & 0 & 1 \\ 0 & 1 & 1 & 1 & 1 & 1 & 1 & 1 & 2 & 2 \\ 0 & 1 & 1 & 2 & 0 & 0 & 1 & 2 & 1 & 0 \\ 0 & 3 & 2 & 2 & 1 & 1 & 1 & 1 & 0 & 0 \end{bmatrix}$$

The problem was solved with the following parameter values: $\alpha = 3$, $q_1 = 0.9$, $q_2 = 0.95$, $h_0 = 1$. The starting point was taken as $x_0 = (0, 0, 0, 0, 1)$, with $f(x_0) = 80$. Table 2 contains the values of iterates and their function

**Table 2**

| Iteration | $x^{(1)}$ | $x^{(2)}$ | $x^{(3)}$ | $x^{(4)}$ | $x^{(5)}$ | $f(x)$ |
|---|---|---|---|---|---|---|
| 0 | 0. | 0. | 0. | 0. | 0. | 80 |
| 1 | 0.111958 | 0.223917 | 0.111958 | 0.111958 | 1.111958 | 63.0894 |
| 6 | 1.097278 | 0.788587 | 0.871869 | 0.281589 | 0.804012 | 34.3990 |
| 11 | 1.069281 | 0.712678 | 0.850455 | 0.868581 | 1.102541 | 25.8384 |
| 16 | 1.063638 | 0.913551 | 1.160926 | 1.101065 | 1.073642 | 24.6941 |
| 21 | 1.076656 | 0.990522 | 1.366388 | 0.910783 | 1.116449 | 22.7825 |
| 26 | 1.129564 | 0.989844 | 1.464870 | 0.936742 | 1.115325 | 22.6504 |
| 31 | 1.135010 | 0.973333 | 1.469467 | 0.939472 | 1.116570 | 22.6090 |
| 36 | 1.119750 | 0.981459 | 1.475895 | 0.913994 | 1.126114 | 22.60392 |
| 41 | 1.128545 | 0.978718 | 1.482843 | 0.925396 | 1.123494 | 22.60168 |
| 46 | 1.122855 | 0.979368 | 1.474063 | 0.919058 | 1.124103 | 22.60064 |
| 51 | 1.124632 | 0.979391 | 1.477879 | 0.920667 | 1.124184 | 22.60023 |

values at the corresponding iterations. The point $x^*$ minimizing $f$ is given by $x^{(1)*} = 1.12434$, $x^{(2)*} = 0.97945$, $x^{(3)*} = 1.47770$, $x^{(4)*} = 0.92023$, $x^{(5)*} = 1.12429$, with

$$f(x^*) = \min_{x \in E_n} \max_{i \in \overline{1,m}} \varphi_i(x) = 22.60016.$$

As can be seen in Table 2, after 51 iterations the approximate solution has 6 correct digits with respect to the function value, and at least 3 digits with respect to $x$. This result should be considered good if one takes into account that the function minimized has a two-dimensional ridge.

An investigation of the computational scheme of the algorithm and numerical experience have shown that the method is highly suitable for solving minimax problems, simple to implement and has a sufficiently high speed of convergence. As shown by the experiments, the speed of convergence and the time required for solving a problem depend on the chosen values of $\alpha$ and $q_2$. In theory, any number greater than one can be chosen as the value of the coefficient of space dilation $\alpha$. However, if $\alpha$ is to close to unity then space dilation has little effect, which leads to slow convergence, especially for functions with multi-dimensional ridges. On the other hand, large values of $\alpha$ yield a rapid decrease of the norm of the matrix $B$, and hence also of the steplengths in the original space, resulting in an increase of the number of iterations with no space dilation, which also slows down convergence. It is advisable to choose $\alpha$ in the range from 2 to 4. The parameter $q_2$ can be set equal to one. To choice of $q_2 < 1$ decreases the length of steps uniformly in all directions (while the matrix $B_k$ shortens the steps unevenly in various directions). Choosing the value of $q_2$ much smaller than unity increases the number of "idle" steps (with no space dilation), which slows down convergence. On the other hand, for problems of large dimension it is frequently advisable to choose $q_2 < 1$, since the "average" index of the shortening of steps resulting from the decrease of the norm of $B_k$ equals $1/\sqrt[n]{\alpha}$ and can be too close to unity for large $n$.

The numerical example described above was solved with constant parameters $\alpha = 3$ and $q_1 = 0.9$ for various values of $q_2$ until the minimum value of $f$ equal to 22.60016 was found. For $q_2 = 1$ this required 69 iterations, among them 52 iterations with space dilation $\left(\frac{52}{69} \approx 0.75\right)$; for $q_2 = 0.95$ we had 57 iterations including 41 space dilations $\left(\frac{41}{57} \approx 0.72\right)$; and $q_2 = 0.9$ resulted in 112 iterations including 39 space dilations $\left(\frac{39}{112} \approx 0.35\right)$. A further decrease of $q_2$ is evidently unreasonable.

The experiments indicate that, apparently, for each problem there exists an asymptotically "optimal" value of $q_2$. One can design algorithms that automatically adjust $q_2$ to that optimal value. This can be done by monitoring the ratio of the number of iterations with space dilations to the total number of iterations, and increasing $q_2$ if this ratio drops below a prespecified number, while decreasing $q_2$ whenever this ratio is too close to unity. Similarly one can regulate the value of $\alpha$. We recommend to choose the value of $q_1$ in the range $0.9 \div 1$.

The algorithm proposed should be applied to problems for which a high accuracy of solutions is required. It needs storing the matrix $B$ of dimension $n \times n$. Due to storage limitations this algorithm can solve minimax problems with relatively few variables (for instance, the internal memory of the BESM-6 computer allows for solving problems with up to 150 variables). Difficulties in storing the matrix $B_k$ arise for large scale problems. In such problems the minimax is usually attained for equal values of several, say $m_1$, functions, with $m_1$ being much smaller than $n$ (the dimension of the problem). Therefore we propose below another algorithm, which avoids storing the matrix and shortens the time of computations.

Let us describe the second algorithm in detail. As in the first method, we start from an arbitrary point $x_0 \in E_n$ and set $\tilde{g}_0 = (0, 0, \ldots, 0) \in E_n$. Suppose that the first $k$ iterations $(k = 1, 2, \ldots)$ resulted in points $x_1$, $x_2, \ldots, x_k$, a vector $\tilde{g}_k$ and a sequence of vectors $\xi_1, \xi_2, \ldots, \xi_k$, where $\xi_v \in E_n (v = 1, 2, \ldots, k)$ are vectors of unit length determining directions in which the space $E_n$ is dilated at each iteration.

To find the point $x_{k+1}$ one has to compute:

1) $f(x_k) = \max\limits_{i} \varphi_i(x_k)$ — the function value at the point $x_k$;

2) $g_f(x_k) = g_{\varphi_{i*}}(x_k)$ — the generalized gradient of $f$ at $x_k$, where $\varphi_{i*}(x_k) = f(x_k)$;

3) $\tilde{g}_k = R_\beta(\xi_k) R_\beta(\xi_{k-1}), \ldots, R_\beta(\xi_1) g_f(x_k)$ — the generalized gradient of the function in the transformed space.

In this algorithm one stores, instead of the matrix $B_k$, the sequence of normalized vectors $\xi_1, \ldots, \xi_k$. The generalized gradient $\tilde{g}_k$, computed at the point $y_k = A_k x_k$, is obtained by applying the operator of space dilation $R_\beta(\xi_1)$ to the vector $g_f(x_k)$, then the dilation operator $R_\beta(\xi_2)$ transforms the vector $R_\beta(\xi_1) g_f(x_k)$ and so on until, finally, the operator $R_\beta(\xi_k)$ of space dilation in the direction $\xi_k$ with the coefficient $\beta = 1/\alpha$ transforms the vector resulting from the sequential application of the operators of space dilation in the directions $\xi_1, \ldots, \xi_{k-1}$.

The formula for $\bar{g}_k$ is obtained from the formula for the generalized gradient when the space $E_n$ is dilated sequentially in the directions $\xi_1, \xi_2, \ldots, \xi_k$:

$$y_k = R_\alpha(\xi_k) R_\alpha(\xi_{k-1}), \ldots, R_\alpha(\xi_1) x_k$$

and from the condition $\bar{g}_k = g_{\psi_k}(y_k)$, where $\psi_k(y_k) = f(A_k^{-1} y_k)$,

$$\psi_k(y_k) = f[R_\alpha(\xi_1) R_\alpha(\xi_2), \ldots, R_\alpha(\xi_k) y_k].$$

Next, we compute:

4) $r_k = \bar{g}_k - \tilde{g}_k$ — the difference of the two generalized gradients;
5) $\beta_k = \|r_k\| / \|g_k\|$.
6) We compare $\beta_k$ with a parameter $q_1$:
   a) if $\beta_k \leqq q_1$ then

   $$x_{k+1} = x_k - h_k R_\beta(\xi_1), \ldots, R_\beta(\xi_k) \frac{\tilde{g}_k}{\|\tilde{g}_k\|},$$

   i.e. the algorithm moves in the same direction and with the same stepsize as at the $k$-th iteration, and we pass to the $(k+2)$-nd iteration;
   b) if $\beta_k > q_1$ then we execute the following steps:
7) we calculate and store the vector $\xi_{k+1} = r_k / \|r_k\|$;
8) we compute $\tilde{g}_{k+1} = R_\beta(\xi_{k+1}) R_\beta(\xi_k), \ldots, R_\beta(\xi_1) g_f(x_k)$.

Taking into account the formula for the generalized gradient $\bar{g}_k$, we finally get

$$\tilde{g}_{k+1} = R_\beta(\xi_{k+1}) \bar{g}_k,$$

where $\tilde{g}_{k+1}$ is the generalized gradient of the function $\varphi_{k+1}$ at the point $\tilde{y}_{k+1}$ which corresponds to $x_k$ after the space was dilated in the direction $\xi_{k+1}$ with the coefficient $\alpha$ at the $(k+1)$-st iteration.

Then the algorithm proceeds as follows:
9) we calculate $h_{k+1} = h_k q_2$, where $0 < q_2 \leqq 1$;
10) we determine the next point $x_{k+1}$ which is obtained by moving in the transformed space along the direction $- \tilde{g}_{k+1}$ from the point $\tilde{y}_{k+1} = R_\alpha(\xi_{k+1}) R_\alpha(\xi_k), \ldots, R_\alpha(\xi_1) x_k$ to the point $y_{k+1} = \tilde{y}_{k+1} - h_{k+1} \tilde{g}_{k+1} / \|\tilde{g}_{k+1}\|$. As indicated above, $y_{k+1} = R_\alpha(\xi_{k+1}), \ldots, R_\alpha(\xi_1) x_{k+1}$, hence $x_{k+1} = R_\beta(\xi_1) R_\beta(\xi_2), \ldots, R_\beta(\xi_{k+1}) y_{k+1}$. From these relations we obtain the formula for $x_{k+1}$:

$$x_{k+1} = x_k - h_k R_\beta(\xi_1) R_\beta(\xi_2), \ldots, R_\beta(\xi_{k+1}) \frac{\tilde{g}_{k+1}}{\|\tilde{g}_{k+1}\|}.$$

Then we pass to the $(k+2)$-nd iteration.

Since the memory for storing the vectors $\xi_1, \ldots, \xi_k$ is limited, one has to stop the above-described algorithm after a certain number $r$ of iterations

$(r < n)$. Then the matrix $B$ is reinitialized, $x_r$ is taken as the new starting point for the next cycle of the method, and the initial stepsize is taken $2-3$ times smaller than the stepsize used at the beginning of the previous cycle. This procedure continues until either sufficiently small function values are attained or the decrease of the function value in some cycle of computations becomes sufficiently small.

In effect, the second method is intermediate between the subgradient algorithm and the algorithm with space dilation. This method was used for solving numerous problems connected with the problem of elaborating optimal standards. The statement of the problem, initial data and procedures for function and gradient evaluations were kindly supplied to us by V. A. Kovalevski and M. I. Shlezinger. This complex max-min problem had 1200 variables and 1536 functions from which the pointwise minima were taken.

For comparing the efficiency of algorithms this problem was also solved by the ordinary subgradient algorithm with no space dilation using the following three rules for stepsize selection:

1) $h_{k+1} = h_k \cdot 0.95,$ $\qquad x_{k+1} = x_k + h_{k+1} \dfrac{g_f(x_k)}{\| g_f(x_k) \|};$

2) $h_{k+1} = h_k \cdot 0.95,$ $\qquad x_{k+1} = x_k + h_{k+1} \, g_f(x_k);$

3) $h_{k+1} = \dfrac{f(x^*) - f(x_k)}{\| g_f(x_k) \|},$ $\quad x_{k+1} = x_k + h_{k+1} \dfrac{g_f(x_k)}{\| g_f(x_k) \|},$

where $f(x^*)$ was a predicted value of the maximum. The best results were obtained with the first stepsize rule (Table 3).

**Table 3**

| Iteration number | Function value | |
|---|---|---|
| | The subgradient algorithm with the first stepsize rule | The second algorithm |
| 1 | $-30.92$ | $-30.92$ |
| 5 | $-\ 7.009$ | $-11.26$ |
| 10 | $-\ 4.548$ | $-\ 8.703$ |
| 15 | $-\ 0.3739$ | $1.143$ |
| 20 | $1.456$ | $2.785$ |
| 25 | $1.911$ | $3.965$ |
| 30 | $1.911$ | $4.551$ |
| 35 | $2.315$ | $4.870$ |
| 40 | $3.350$ | $5.010$ |
| 45 | $3.403$ | $5.197$ |
| 50 | $3.826$ | $5.277$ |
| 55 | $3.779$ | $5.312$ |
| 60 | $4.025$ | $5.333$ |
| 65 | $3.732$ | $5.347$ |
| 71 | $3.882$ | |

For solving this problem with the second algorithm described above, the application of which was warranted by the considerable dimension of the problem ($n = 1200$), a program in FORTRAN was written for the BESM-6 computer. The value of $\alpha$ was set equal to 3.

An analysis of the results, obtained by both methods, presented in Table 3, shows that for attaining comparable function values the second algorithm required significantly fewer iterations and less computation time than the subgradient algorithm with no space dilation. For instance, an acceptable function value of 3.88 was obtained by the subgradient algorithm with no space dilation at the 71-st iteration, whereas the second modification of the algorithm obtained the value of $f$ equal to 3.96 already at the 25-th iteration. Moreover, the second algorithm found the value of the maximized function (5.07 at the 39-th iteration) which very accurately agrees with the theoretical maximum. Such accuracy could not be attained by the subgradient algorithm (it required too many iterations and a long time of computation).

## 4.6 Application of Methods for Minimizing Nonsmooth Functions to Problems of Interpreting Gravimetric Observations

This section describes the research conducted by the author together with I. G. Ovrutski [63].

The broad introduction of computers to the practice of geophysical explorations makes it possible to use new methods for geological interpretation of gravitational anomalies. When constructing such methods one should always take account of the fact that the inverse problem of gravitational prospecting belongs to the class of ill-posed problems of mathematical physics and so its solution requires special mathematical models and methods [37, 99].

We shall consider the methodology of solving the inverse problem of gravitational prospecting by reducing it to a nonlinear constrained minimax problem, which is then solved by the generalized gradient method with space dilation, namely the $r$-algorithm.

Let $x = \{x^{(1)}, \ldots, x^{(n)}\} \in E_n$ denote the vector of parameters of the model of the geological environment, let $F_i$ denote the solution operator of the direct problem of gravitational exploration, which for a given $x \in E_n$ yields the theoretical values of anomalies of the gravity force at the $i$-th observation point, and let $\Delta g_i$ denote the measured values of gravitational anomalies at point $i$. The inverse problem of gravitational prospecting consists in finding a vector $x \in E_n$ that minimizes, subject to certain constraints, the objective function characterizing the deviations of the theoretical values of gravitational anomalies from the measured ones.

In general, the solution of the inverse problem is nonunique for a given objective function, since for any measured anomaly one can always find several equivalent distributions of "interfering masses". All the same, gravitational prospecting is successfully used for solving geological problems. The explanation is that the interpreting geophysicist makes use of additional geological information about the form and quantity of the anomalous bodies and the density profile of the geological section, and data obtained by other geophysical methods, so that the search for a solution can be conducted within the framework of a finite-dimensional parametric model with a priori constrained parameter values. In geophysics the following two objective functions usually serve as criteria of quality of interpretation:

$$Q_1 = \frac{1}{m} \sum_{i=1}^{m} c_i (F_i(x) - \Delta g_i)^2 \tag{4.172}$$

for estimating parameters by the least-squares method, and

$$Q_2 = \max_i |c_i (F_i(x) - \Delta g_i)|, \tag{4.173}$$

for estimating parameters optimal in the Chebyshev sense (with respect to the maximum deviation), where $c_i$ denote weighing factors. The choice of the objective function is dictated by a specific geological situations as well as by information about measurement and modelling errors.

Since measurements of anomalies of the gravity force always contain random instrumental errors and the model of the geological environment is simplified, the deviation of the observed anomaly from the theoretical one obtained by applying the operator $F$ to the parameter vector $x$ is a certain combination of the instrumental and modelling errors. Under such conditions estimating the parameters by the least squares method has no clear advantage over estimating by means of the Chebyshev criterion (in most practical cases the error due to using only a finite number of parameters is much larger than the error resulting from imprecise measurements).

We shall consider here an algorithm for solving the inverse problem with the Chebyshev criterion. Similar algorithms can also be used for the least squares method; however, as will be seen later, the interpretation with the Chebyshev criterion requires, as a rule, less computational effort. Thus we want to choose the model parameters $x$ so as to minimize the maximum deviation of the theoretical anomaly from the observed one. In practice it suffices to decrease the deviation below a certain number $\varepsilon$, which is chosen by the geologist on the basis of the accuracy of the experimental data and his assumptions on the modelling errors.

The inverse problem of gravitational prospecting can be formulated as a minimax problem of the following form: minimize

$$Q = \max_i |F_i(x) - \Delta g_i|, \tag{4.174}$$

subject to the constraints

$$f_k(x) \leqq 0, \quad k = 1, \ldots, p. \tag{4.175}$$

Typical constraints for the problem in question are of the form $x_a \leqq x \leqq x_b$, where $x_a$ and $x_b$ are the left and right bounds on the parameter values, respectively.

The solution of problem (4.174)−(4.175) can be reduced to the solution of the following unconstrained minimax problem of minimizing

$$Q_\lambda(x) = \max_i |F_i(x) - \Delta g_i| + \lambda \sum_{k=1}^{p} f_k(x)\, \delta_k(x), \tag{4.176}$$

where the second term of the right side of (4.176) is the penalty function,

$$\delta_k(x) = \begin{cases} 0 & \text{if } f_k(x) \leqq 0, \\ 1 & \text{if } f_k(x) > 0, \end{cases}$$

and $\lambda$ is a sufficiently large penalty coefficient [22]. The function $Q_\lambda$ is nonsmooth. Consequently, to minimize $Q_\lambda$ one should use generalized gradient methods, which have been specially constructed for minimizing nonsmooth functions.

The form of the operator $F$ depends entirely on the way of modelling of the geological environment. If the geological environment is approximated by three-dimensional rectangular prisms, then the law of gravity yields the following well-known formula [100]

$$F_i(x) = -f \sum_{j=1}^{n} \sigma_j \left|\,\left|\,\left|\, (\eta_j - x_i^{(2)}) \ln\left[(\xi_j - x_i^{(1)}) + R_{ji}\right]\right.\right.\right.$$

$$+ (\xi_j - x_i^{(1)}) \ln\left[(\eta_j - x_i^{(2)}) + R_{ji}\right]$$

$$+ (x_i^{(3)} - \zeta_j)\, \text{arc tg}\, \frac{(\xi_j - x_i^{(1)})(\eta_j - x_i^{(2)})}{(\zeta_j - x_i^{(3)})\, R_{ji}} \left.\left.\left.\vphantom{\frac{a}{b}}\right|_{\xi_{j1}}^{\xi_{j2}}\right|_{\eta_{j1}}^{\eta_{j2}}\right|_{\zeta_{j1}}^{\zeta_{j2}}, \tag{4.177}$$

where $f$ is the gravitational constant, $x_i^{(1)}$, $x_i^{(2)}$ and $x_i^{(3)}$ are the coordinates of point $i$, $\xi_{j1}$, $\xi_{j2}$, $\eta_{j1}$, $\eta_{j2}$, $\zeta_{j1}$ and $\zeta_{j2}$ are the axis coordinates of the $j$-th prism, $\sigma_j$ is the surplus density, and

$$R_{ji} = \sqrt{(\xi_j - x_i^{(1)})^2 + (\eta_j - x_i^{(2)})^2 + (\zeta_j - x_i^{(3)})^2}.$$

It is difficult to minimize the function given by (4.176), since by its physical nature $Q_\lambda$ is a nonsmooth and gully-type function. This can be explained by the fact that the values of $\Delta g_i$ are influenced by heterogeneous factors. Thus the upper edges $\zeta_{j1}$ of the bodies influence the observed gravitational field stronger than the lower edges $\zeta_{j2}$, and hence in the multidimensional space the level sets of the function are stretched out, which impedes minimization.

For solves such a class of problems we propose to use the modified generalized gradient descent with space dilation along the direction of the difference of two successive gradients. The computation of the gradient (or its analogue) of the max function $\varphi(x) = \max_i f_i(x)$ at a point $x_0$ requires calculating $\max_i f_i(x_0)$ and computing the gradient of a function $f_i$ for which the maximum is attained.

Most calculations in gradient-type algorithms are, as a rule, connected with gradient evaluations. To simplify the programming, for complicated functions the gradient is computed by finite differences. Let us compare the times of gradient evaluations for functions of the forms (4.172) and (4.173). As a unit of computational time we shall adopt the time required for evaluating $F_i$ at one observation point for one body of the form of a rectangular prism. Suppose that the model consists of $n$ bodies, the measurements were collected at $m$ points, and the vector $x$ of unknown parameters is of dimension $t$. Then the expenditure of computer time equals $mn + t$ for calculating the gradient of function (4.173) by finite differences, and $m(t + n)$, i.e. is larger by the factor of $(t/n + 1)/(1 + t/mn)$, for an analogous computation of the gradient of function (4.172). For instance, with $n = 1$, $m = 20$ and $t = 5$ each gradient evaluation of function (4.172) is about five times more time-consuming than the evaluation of the gradient of function (4.173). In real-life problems the number $t$ of parameters may be in the tens. Therefore if the gradient method for minimizing function (4.173) is not significantly slower than the method for minimizing function (4.172), then minimizing (4.173) up to a given accuracy will require much less time than minimizing (4.172) with a comparable accuracy. This motivates our interest in the construction of algorithms for interpreting gravitational anomalies according to the Chebyshev criterion.

For solving the inverse problem of gravitational prospecting by the algorithm proposed we elaborated a program in FORTRAN-IV for the "Minsk-32" computer. The program consists of a master segment, a segment for evaluating the objective function $Q_\lambda$ and a finite-difference analogue of the generalized gradient, and a segment for minimization. The master segment organizes the input of initial data, chooses parameters of calculations, prints intermediate and final results, and organizes the exit from the program. In the segment for evaluating the objective function $Q_\lambda$ are programs which calculate by formulae (4.176) and (4.177) the solution to the direct problem for a model of the geological invironment approximated by a set of three-dimensional rectangular prisms, and programs for computing an analogue of the generalized gradient by the method of finite differences. The minimizing segment implements the algorithm of generalized gradient descent with space dilation along the direction of the difference of two successive gradients (the $r$-algorithm), described in Sect. 4.5. The program was tested on simulated and practical problems [63].

The validation of our methodology on models enabled us to estimate the areas of possible applications and the solution accuracy. The proposed methodology for interpreting anomalies of the gravity force may be successfully used for determining the morphology of intrusions if density characteristics of the environment are known, for determining the position of contact gravitating surfaces and vertical contacts of bodies, for estimating the lower boundary of the distribution of interfering masses, etc.

## 4.7 Other Areas of Applications of Generalized Gradient Methods

The preceding sections described the use of generalized gradient methods for implementing the schemes of decomposition with respect to constraints or variables and certain types of parametric decomposition, for solving production-transportation type problems of operative planning, and for solving minimax problems and problems of measurement data processing. These examples in no way exhaust possible applications of the generalized gradient methods. In this section we shall briefly review other application areas.

### The Solution of Optimal Design Problems

Mathematical models arising in the problems of optimal design of complex dynamic processes, machines, buildings and structures are of the form of a nonlinear programming problem with inequality and equality constraints, having both discrete and continuous variables. The problem frequently turns out to be multi-extremal. Usually, to solve problems of optimal design generalized gradient methods are used for finding local optima with respect to the continuous variables together with combinatorial schemes for varying the discrete variables and combinatorial or stochastic techniques for finding initial approximations. Constraints are treated by using either reduction methods for eliminating the variables involved in equality constraints, or the method of nonsmooth penalty functions [22, 9, 28] for mixed, equality and inequality constraints. Specific applications of methods for minimizing nonsmooth functions to design problems are described in [58].

Let us consider in detail the method of nonsmooth penalty functions. Following [9], for the convex programming problem

$$\min f_0(x), \quad \text{s.t.} f_i(x) \leqq 0, \quad i = 1, \ldots, m, \tag{4.178}$$

let us define the function

$$S(x) = f_0(x) + \sum_{i=1}^{m} p_i[f_i(x)], \tag{4.179}$$

where $p_i$, $i = 1, \ldots, m$, are convex functions satisfying $p_i(t) = 0$ for $t \leqq 0$, $p_i(t) > 0$ for $t > 0$. Suppose that there exists a point $x^*$ which minimizes $S(x)$ over all $x$.

**Theorem 4.2.** *For $x^*$ to be a solution of the original problem it is necessary that*

$$\lim_{t \to 0+} \frac{p_i(t)}{t} \geqq \bar{y}_i, \quad i = 1, \ldots, m,$$

*where $\bar{y} = (\bar{y}_1, \ldots, \bar{y}_m)$ is a Lagrange multiplier vector of problem* (4.178). *Moreover, if $\lim_{t \to 0+} p_i(t)/t > \bar{y}_i$ then the sets of minimum points of* (4.178) *and* (4.179) *are equal.*

We should add that nonsmooth penalty functions were first seriously investigated by I. I. Eremin [22]. The simplest version of a nonsmooth penalty function is of the form

$$p(t) = \begin{cases} 0 & \text{if } t \leqq 0, \\ c\,t & \text{if } t > 0. \end{cases} \tag{4.180}$$

It follows from Theorem 4.2 that to ensure that problem (4.178) is equivalent to the problem of minimizing $S$ with

$$p_i(t) = \begin{cases} 0 & \text{if } t \leqq 0, \\ c_i\,t & \text{if } t > 0, \ i = 1, \ldots, m, \end{cases} \tag{4.181}$$

it suffices to take $c_i > \bar{y}_i$.

Of much interest is the penalty function

$$T(x) = f_0(x) + p\left[\max_{i \in \overline{1,m}} f_i(x)\right],$$

where $p$ is defined by formula (4.180). It is easy to see that if $c > \sum_{i=1}^{m} \bar{y}_i$ then

$$p(\max_{i \in \overline{1,m}} f_i(x)) \geqq \sum_{i=1}^{m} p_i[f_i(x)],$$

where $p_i(t)$ are defined by (4.181), $c = \sum_{i=1}^{m} c_i$, and $c_i > \bar{y}_i$. Thus for $c > \sum_{i=1}^{m} \bar{y}_i$ the problem of minimizing $T$ is equivalent to problem (4.178).

The method of nonsmooth penalty functions can be successfully used for finding local extrema in nonlinear programming problems of the following general form:
minimize:

$$f_0(x), \tag{4.182}$$

subject to the constraints

$$\varphi_i(x) \leqq 0, \quad i = 1, \ldots, m, \quad \psi_j(x) = 0, \quad j = 1, \ldots, l, \tag{4.183}$$

by converting them to the problem of minimizing the function

$$S(x) = f_0(x) + \sum_{i=1}^{m} c_i\,\varphi_i^+(x) + \sum_{j=1}^{l} d_j\,|\psi_j(x)|. \tag{4.184}$$

If the Lagrange multipliers of problem $(4.182)-(4.183)$ exist and the values of $c_i$ and $d_j$ are sufficiently large, then the problem of minimizing $S$ is equivalent to problem $(4.182)-(4.183)$.

The application of nonsmooth penalty functions usually leads to problems of minimizing gully-type functions. For solving them we recommend the algorithms with space dilation in the direction of the difference of two successive gradients.

**Problems of Discrete and Mixed Discrete-Continuous Programming**

Problems of discrete and discrete-continuous types are frequently solved by the branch and bound method. The efficiency of this method depends strongly on the complexity of finding estimates of the optimum function value and on their accuracy. In recent years much attention has been devoted to the dual approach to finding such estimates. Its essence consists in the following [32].

Consider the nonlinear programming problem

$$\min_{x \in X \subset E_n} f_0(x) \tag{4.185}$$

subject to the constraints

$$f_i(x) \le 0, \quad i = 1, \ldots, m, \tag{4.186}$$

where the set $X$ is compact and partially discrete. Let us introduce the Lagrange function

$$L(x, u) = f_0(x) + \sum_{i=1}^{m} u_i f_i(x),$$

and let

$$\Phi(u) = \min_{x \in X} \left[ f_0(x) + \sum_{i=1}^{m} u_i f_i(x) \right]. \tag{4.187}$$

Then the function $\Phi$ is concave on its domain. In particular, if $X$ is a finite discrete set then $\Phi$ is a piecewise linear concave function. Let $Q = \max_{u \ge 0} \Phi(u)$.

**Theorem 4.3.** *If $f^*$ denotes the optimum value of problem $(4.185)-(4.186)$ then $Q \le f^*$.*

*Proof.* Let $x^*$ denote a solution to problem $(4.185)-(4.186)$. Then $f_i(x^*) \le 0$, $i = 1, \ldots, m$, and $f_0(x^*) = f^*$. For any $u \ge 0$

$$\Phi(u) \le L(x^*, u) \le f^*.$$

Therefore $Q = \max_{u \ge 0} \Phi(u) \le f^*$, as required.  $\square$

Thus $Q$ may serve as a lower estimate of $f^*$.

The problem of finding $\max_{u \ge 0} \Phi(u)$ can be solved by the subgradient algorithm with projection on the set $u \ge 0$ if the value of $m$ is large, or the $r$-algorithm if the value of $m$ does not exceed several hundred. The

complexity of the auxiliary problem (4.187) can be considerable, but in many cases of interest this problem can be solved relatively easily. As an example we shall consider, following [32], the problem of linear integer programming
minimize

$$(c, w),$$

subject to

$$A w = d, \quad w \geqq 0, \tag{4.188}$$

$w$ has integer components,

where $A$ is an $m \times (n + m)$-matrix of rank $m$ with integer entries, $d$ is an integer vector of dimension $m$, and $c$ is an $(n + m)$-vector. Let $B$ be a nonsingular $m \times m$ matrix formed from the columns of $A$, and let the remaining columns form a matrix $N$. If we divide $w$ into subvectors $y$ and $x$ corresponding to $B$ and $N$, and similarly split $c$ into $c_B$ and $c_N$, then problem (4.188) reduces to the following:
minimize

$$[c_B B^{-1} d + (c_n - c_B B^{-1} N) x],$$

subject to the constraints

$$B^{-1} N x \leqq B^{-1} d, \quad B^{-1} N x = B^{-1} d (\text{mod } 1),$$

$$x \geqq 0, \quad x \text{ has integer components.}$$

Introduce the following notation

$$X \quad = \{x: B^{-1} N x = B^{-1} d \ (\text{mod } 1), x \geqq 0, x \text{ is integer}\},$$

$$f(x) = (c_N - c_B B^{-1} N) x;$$

$$g(x) = B^{-1} N x; \quad b = B^{-1} d.$$

Then the dual problem is of the form

$$\min_{x \in X} [f(x) + (u, g(x) - b)]$$

and reduces to the problem of finding the shortest path in a special network comprising $|\det B|$ vertices. An efficient algorithm for this problem can be found in [42].

The dual approach together with the subgradient method was successfully applied to the travelling salesman problem by Held and Karp [93], to integer multicommodity network problems by Held, Wolfe and Crowder [44], and to optimal scheduling problems by Fisher [31].

# Concluding Remarks

The preceding chapters show that the subgradient method is the simplest and most general algorithm among methods for nonsmooth optimization. Unfortunately, its speed of convergence is low and not always easily controllable. As far as algorithms with space dilation are concerned, the most thorough experimental investigation has been carried out for the $r$-algorithm. If the number of variables in the minimization problem does not exceed $200-300$, in most cases the efficiency of using the $r$-algorithms instead of the subgradient method does not raise any doubt. For larger dimensions a considerable part of the computational work is spent on updating matrices of space dilation, and so the $r$-algorithm may require many times more work per iteration than the subgradient method. Thus the $r$-algorithm becomes more suitable if the percentage of time taken by matrix updatings decreases and a higher accuracy of solutions is required. To estimate the solution time for a given problem one can use the following empirical rule for the $r$-algorithms: the relative accuracy with respect to function values increases by an order after every $n \div 1.5\,n$ iterations, where $n$ is the number of variables of the objective function. As far as the SDG method is concerned, its application appears to be reasonable only if the minimum value of the objective function is known; otherwise the $r$-algorithm turns out to be more efficient.

The methods of $\varepsilon$-subgradients are in some way intermediate between the subgradient algorithm and the methods with space dilation. Unfortunately, they have not been sufficiently investigated in practice, and so far theoretical results on their speed of convergence and numerical experiments have not been particularly promising [101].

In the near future one may expect a development of methods for minimizing convex functions which combine the cutting plane approach with space dilation. This could lead to algorithms at least as efficient as the $r$-algorithm, but with a theoretically guaranteed speed of convergence. Especially interesting is the problem of constructing such nonmonotone versions of $\varepsilon$-subgradient methods which would have a significantly higher speed of convergence than the subgradient method and would not require much additional storage.

Of great importance for solving large scale problems by methods of mixed decomposition (with respect to both constraints and variables) are efficiently algorithms for finding saddle points of convex-concave non-smooth functions. This creates the problem of extending gradient algorithms with space transformation to saddle point problems.

Advances in these directions will additionally stimulate the application of generalized gradient methods for minimizing nonsmooth functions to the solution of complex problems of optimal planning, design and operations research.

# References [1]


1    Abadie, J., Williams, A.: Dual and parametric methods in decomposition. In: Recent Advances in Mathematical Programming (R.L. Graves, P. Wolfe, eds.), pp. 149–158. McGraw-Hill: New York 1963

2    Agmon, D.: The relaxation method for linear inequalities. Canad. J. Math. 6, 382–392 (1954)

3    Balinski, M.L., Wolfe, P. (eds.): Nondifferentiable optimization. Math. Programming Stud. 3. North Holland: Amsterdam 1975

4*    Bakaev, A.A., Mikhalevič, V.S., Branovitskaya, S.V., Shor, N.Z.: Methodology and experience in solving large scale network transportation problems on a computer. In: Mathematical Methods and Problems of Production, pp 247–267. Moscow 1963

5*    Bazhenov, L.G.: On the conditions for convergence of methods for minimizing almost differentiable functions. Kibernetika (Kiev), no. 4, 71–72 (1972)

6*    Belaeva, L.V., Zhurbenko, N.G., Shor, N.Z.: On a certain nonlinear transportation problem. In: Teor. Optimal. Rešenii, Trudy Sem. Nauč. Sov. Akad. Nauk Ukrain. SSSR po Kibernet., pp. 83–89. Kiev 1976

7*    Belaeva, L.U., Zhurbenko, N.G., Shor, N.Z.: On a method for solving a class of dynamic distribution problems. Econom. i Mat. Metody 14, 137–146 (1978)

8    Bertsekas, D., Mitter, S.: A descent numerical method for optimization problems with nondifferentiable cost functionals. SIAM J. Control 11, 637–652 (1973)

9    Bertsekas, D.: Necessary and sufficient conditions for a penalty method to be exact. Math. Programming 9, 87–99 (1975)

10    Bertsekas, D.: Nondifferentiable optimization via approximation. In: Math. Programming Stud. 3 (M.L. Balinski, P. Wolfe, eds.), pp. 1–25. North Holland: Amsterdam 1975

11    Best, M.: A method to accelerate the rate of convergence of a class of optimization algorithms. Math. Programming 9, 139–160 (1975)

12    Busemann, H.: Convex surfaces. Interscience: New York 1958

13    Camerini, P., Fratta, L., Maffioli, F.: On improving relaxation methods by modified gradient techniques. In: Math. Programming Stud. 3 (M.L. Balinski, P. Wolfe, eds.), pp. 26–34. North Holland: Amsterdam 1975

14    Clarke, F.H.: Necessary conditions for nonsmooth problems in optimal control and the calculus of variations. Diss. Doctor Philos. Washington 1973

15    Clarke, F.H.: Generalized gradients and applications. Trans. Amer. Math. Soc. 205, 47–262 (1975)

16    Dantzig, G., Wolfe, P.: Decomposition principles for linear programs. Oper. Res. 8, 101–111 (1960)


---

[1]    All asterisked items originally published in Russian.
Additional references not referred to within this volume will be listed on p. 156 ff.

17    Demyanov, V.F.: Extremal basis method in minimax problems. Ž. Vycisl. Mat. i
      Mat. Fiz. 17, 512–517 (1977)
18*   Demyanov, V.F., Malozemov, V.N.: Introduction to minimax. Nauka: Moscow 1972
19    Doob, J.L.: Stochastic processes. Wiley: New York 1983
20    Elzinga, J., Moore, T.: A central cutting plane algorithm for the convex program-
      ming problem. Math. Programming 8, 134–145 (1975)
21*   Eremin, I.I.: Iterative method for Chebyshev approximations of inconsistent
      systems of linear inequalities. Dokl. Akad. Nauk SSSR 143, 1254–1256 (1962)
22*   Eremin, I.I.: On a penalty method in convex programming. Kibernetika (Kiev),
      no. 4, 63–67 (1967)
23*   Eremin, I.I., Astaf'ev, N.N.: Introduction to the theory of linear and convex
      programming. Nauka: Moscow 1976
24*   Ermol'ev, Yu.N.: Methods for solving nonlinear extremal problems. Kibernetika
      (Kiev), no. 4, 1–17 (1966)
25*   Ermol'ev, Yu.N.: Methods of stochastic programming. Nauka: Moscow 1976
26*   Ermol'ev, Yu.N., Shor, N.Z.: On minimization of nondifferentiable functions.
      Kibernetika (Kiev), no. 1, 101–102 (1967)
27*   Ermol'ev, Yu.N., Shor, N.Z.: A random search method for two-stage problems of
      stochastic programming and its generalization. Kibernetika (Kiev), no. 1, 90–92
      (1968)
28    Evans, J., Gould, F., Tolle, J.: Exact penalty functions in nonlinear programming.
      Math. Programming 4, 72–97 (1973)
29*   Faddeev, D.K., Faddeeva, V.N.: Computational methods of linear algebra. Fiz-
      matgiz: Moscow 1960
30    Fiacco, A., Mc Cormick, G.: Nonlinear programming: Sequential unconstrained
      minimization techniques. Wiley: New York 1968
31    Fisher, N.: Optimal solution of scheduling problems using Lagrange multipliers.
      Part 1. Oper. Res. 21, 1114–1127 (1973)
32    Fisher, N., Northup, W., Shapiro, J.: Using duality to solve discrete optimization
      problems. In: Math. Programming Stud. 3 (M. L. Balinski, P. Wolfe, eds.), pp
      56–94. North Holland: Amsterdam 1975
33*   Gantmakher, F.R.: Theory of matrices. Gostekhizdat: Moscow 1953
34*   Gel'fand, I.N., Tsetlin, M.L.: The principle of nonlocal search in the problems of
      automatic optimization. Dokl. Akad. Nauk SSSR 137, 295–298 (1961)
35*   Gershovich, V.I.: On the experience in solving the problem of coal distribution for
      coking. In: Teor. Optimal. Resenii, Trudy Sem. Nauc. Sov. Akad. Nauk Ukrain
      SSSR po Kibernet. pp. 22–35. Kiev 1977
36    Gerstenhaber, M.: Solution of large scale transportation problems. In: Combina-
      torial analysis (R. Bellman, M. Hall, eds.), pp. 251–260. Amer. Math. Soc.:
      Providence 1960
37*   Glasko, V.B., Gushchin, G.V., Starostenko, V.I.: On an application of the
      regularization method of A.N. Tikhonov to the solution of systems of nonlinear
      equations. Ž. Vycisl. Mat. i Mat. Fiz. 16, 283–292 (1976)
38    Goldstein A.: Optimization with corners. In: Nonlinear Programming 3 (O.L.
      Mangasarian, R.R. Meyer, S.M. Robinson, eds.), pp. 215–230. Academic Press:
      New York 1975
39*   Gol'shtein, E.G.: Generalized gradient method for sadde point seeking. Ekonom. i
      Mat. Metody 8, 569–579 (1972)
40*   Gol'shtein, E.G., Tret'yakov, N.V.: A gradient method for minimization and
      algorithms for convex programming related to modified Lagrangian functions.
      Ekonom. i Mat. Metody 11, 730–742 (1975)
41*   Gol'shtein, E.G., Yudin, D.B.: New directions in linear programming. Sovetskoe
      Radio: Moscow 1966
42    Gorry, G., Northup, W., Shapiro, J.: Computational experience with a group
      integer programming algorithm. Math. Programming 4, 171–192 (1973)

43   Held, M., Carp, R.: The travelling salesman problem and minimum spanning trees.
     Part 2. Math. Programming 1, 6–25 (1971)
44   Held, M., Wolfe, P., Crowder, H.: Validation of subgradient optimization. Math.
     Programming 6, 62–88 (1974)
45   Huang, H.Y.: Unified approach to quadratically convergent algorithms for function
     minimization. J. Optim. Theory Appl. 5, 402–423 (1970)
46*  Kantorovich, L.V.: On the method of steepest descent. Dokl. Akad. Nauk SSSR 56,
     233–236 (1947)
47   Kantorovich, L.V.: Economic calculation of the best utilization of resources.
     Izdatel'stwo Akad. Nauk SSSR: Moscow 1959
48   Kelley, J.: The cutting plane method for solving convex programs. J. Soc. Ind.
     Appl. Math. 8, 703–712 (1960)
49*  Krasnosel'skii, M.A.: Topological methods in the theory of nonlinear integral
     equations. Gostekhizdat: Moscow 1956
50   Künzi, H.P., Krelle, W.: Nonlinear programming. Blaisdell: Waltham 1966
51   Lemaréchal, C.: An algorithm for minimizing convex functions. In: Proceedings of
     the IFIP Congress, pp. 552–556. North Holland: Amsterdam 1974
52   Lemaréchal, C.: An extension of Davidon methods to nondifferentiable prob-
     lems. In: Math. Programming Stud. 3 (M.L. Balinski, P. Wolfe, eds.), pp. 95–100.
     North Holland: Amsterdam 1975
53*  Levin, A.Yu.: On an algorithm for minimizing convex functions. Dokl. Akad. Nauk
     SSSR 160, 1244–1247 (1965)
54*  Lyubich, Yu.I., Maistrovskii, G.D.: General theory of relaxation processes for
     convex functions. Uspehi Mat. Nauk 1, 57–112 (1970)
55   Mifflin, R.: An algorithm for constrained optimization with semismooth functions.
     Math. Oper. Res. 2, 191–207 (1977)
56   Mifflin, R.: Semismooth and semiconvex functions in constrained optimization.
     SIAM J. Control Optim. 15, 959–972 (1977)
57*  Mikhalevich, V.S., Ermol'ev, Yu.M., Shkurba, V.V., Shor, N.Z.: Complex systems
     and the solution of extremal problems. Kibernetika (Kiev), no. 5, 29–39 (1967)
58   Mikhalevich, V.S., Shor, N.Z., Galustova, L.A. (eds.): Computational methods for
     choosing optimal design decisions. Naukova Dumka: Kiev 1977
59   Moiseev, N.N.: Numerical methods in the theory of optimal systems. Nauka:
     Moscow 1971
60   Motzkin, T., Shoenberg, I.: The relaxation method for linear inequalities. Canad. J.
     Math. 6, 393–404 (1954)
61*  Nurminski, E.A.: A quasi-gradient method for solving the nonlinear programming
     problem. Kibernetika (Kiev) no. 1, 122–125 (1973)
62*  Nurminski, E.A.: On the continuity of $\varepsilon$-subgradient mappings. Kibernetika (Kiev),
     no. 5, 148–149 (1977)
63*  Ovrutskii, I.G., Shor, N.Z.: Application of methods for minimizing nonsmooth
     functions to the solution of the problem of gravimetric data interpretation. Kiber-
     netika (Kiev), no. 2, 57–64 (1976)
64*  Polyak, B.T.: A general method for solving extremal problems. Dokl. Akad. Nauk
     SSSR 174, 33–36 (1967)
65*  Polyak, B.T.: Minimization of nonsmooth functionals. Z. Vycisl. Mat. i Mat. Fiz. 9,
     509–521 (1969)
66*  Polyak, B.T.: The ethod of conjugate gradients. Trudy II Zimnei Skoly po Mat.
     Programmirovaniyu i Smeznim Vopr., wyp. 1, pp. 152–201, 1969
67*  Primak, M.E.: On convergence of a modified method of Chebyshev's centers for
     solving the convex programming problem. Kibernetika (Kiev), no. 5, 100–102
     (1977)
68*  Pshenichny, B.N.: Necessary conditions for an extremum. Nauka: Moscow 1969
69*  Pshenichny, B.N., Danilin, Yu.M.: Numerical methods in extremal problems.
     Nauka: Moscow 1975

70   Rockafellar, R.T.: Convex analysis. Princeton University Press: Princeton 1970
71   Rosen, J.B.: Convex partition programming. In: Recent advances in mathematical programming (R.L. Graves, P. Wolfe, eds.), pp. 159–176, McGraw Hill: New York 1963
72*  Shabashova, L.P.: Gradient methods for solving nonlinear minimax problems. Diss. Doctor Philos. Dnépropetrowsk 1973
73*  Shepilov, M.A.: On gradient and penalty methods in mathematical programming problems. Diss. Doctor Philos. Moscow 1974
74*  Shepilov, M.A.: On a method of generalized gradient for finding the absolute minimum of a convex function. Kibernetika (Kiev), no. 4, 52–57 (1976)
75*  Shor, N.Z.: An application of the method of gradient descent to the solution of the network transportation problem. In: Materialy Naucnovo Seminara po Teoret i Priklad. Voprosam Kibernet. i Issted. Operacii, Nucnyi Sov. po Kibernet. Akad. Nauk Ukrain. SSSR, vyp. 1, pp. 9–17, Kiev 1962
76*  Shor, N.Z.: On the structure of algorithms for numerical solution of problems of optimal planning and design. Diss. Doctor Philos. Kiev 1964
77*  Shor, N.Z.: An application of the generalized gradient descent in block programming. Kibernetika (Kiev), no. 3, 53–55 (1967)
78*  Shor, N.Z.: Multistage convex stochastic programming. In: Teor. Optimal Resenii, Trudy Sem. Nauc. Sov. Akad. Nauk Ukrain. SSR po Kibernet., pp. 48–58. Kiev 1967
79*  Shor, N.Z.: On the speed of convergence of the generalized gradient descent. Kibernetika (Kiev), no. 3, 98–99 (1968)
80*  Shor, N.Z.: The generalized gradient descent. In: Trudy I Zimnei Skoly po Mat. Programmirovaniyu, vyp. 3, pp. 578–585, 1969
81*  Shor, N.Z.: On the speed of convergence of the method of generalized gradient descent with space dilation. Kibernetika (Kiev), no. 2, 80–85 (1970)
82*  Shor, N.Z.: Methods for minimizing nondifferentiable functions and their applications. Dis. Doctor Sci. Kiev 1970
83*  Shor, N.Z.: An application of the operation of space dilation to the problems of minimizing convex functions. Kibernetika (Kiev), no. 1, 6–12 (1970)
84*  Shor, N.Z.: On a method for minimizing almost differentiable functions. Kibernetika (Kiev), no. 4, 65–70 (1972)
85*  Shor, N.Z.: The analysis of convergence of a gradient type method with space dilation in the direction of the difference of two successive gradients. Kibernetika (Kiev), no. 4, 48–53 (1975)
86*  Shor, N.Z.: Generalized gradient methods for minimizing nonsmooth functions and their application to mathematical programming problems (Review). Ekon. i Mat. Metody, 12, 337–356 (1976)
87*  Shor, N.Z.: A method of section with space dilation for solving convex programming problems. Kibernetika (Kiev), no. 1, 94–95 (1977)
88*  Shor, N.Z., Biletskii, V.I.: A method of space dilation for accelerating convergence in gully-type problems. In: Teor. Optimal. Rešenii, Trudy Sem. Nauč. Sov. Akad. Nauk Ukrain. SSR po Kibernet. no. 2, pp. 3–18, Kiev 1969
89*  Shor, N.Z., Galustova, L.A., Momot, A.I.: An application of mathematical methods to the optimal design of the central gas supply system, accounting for the dynamics of its development. Kibernetika (Kiev), no. 1, 69–74 (1978)
90*  Shor, N.Z., Gorbach, G.I.: The solution of distribution type problems by the method of generalized gradient descent. In: Teor. Optimal. Rešenii, Trudy Sem. Nauč. Sov. Akad. Nauk Ukrain. SSR po Kibernet., no. 1, pp. 59–71, Kiev 1967
91*  Shor, N.Z., Ivanova, L.V.: On a certain iterative method for solving linear programming problems and matrix games. In: Teor. Optimal. Rešenii, Trudy Sem. Nauč. Sov. Akad. Nauk Ukrain. SSR, po Kibernet. no. 3, pp. 22–30, Kiev 1969
92*  Shor, N.Z., Rosina, N.I.: Scheme for partitioning linear and convex programming problems and its application to the solution of transportation planning problems.

In: Dokl. I. Vsesoyuz. Konf. po Optimizacii i Modelirovaniyu Transport. Setei, pp. 225–237, Kiev 1967

93*  Shor, N.Z., Shabashova, L.I.: On the solution of minimax problems by the method of generalized gradient descent with space dilation. Kibernetika (Kiev), no. 1, 82–88 (1972)

94*  Shor, N.Z., Shchepakin, M.B.: An algorithm for solving the two-stage problem of stochastic programming. Kibernetika (Kiev), no. 3, 56–58 (1968)

95*  Shor, N.Z., Zhurbenko, N.G.: A minimization method using the operation of space dilation in the direction of the difference of two successive gradients. Kibernetika (Kiev), no. 3, 51–59 (1971)

96*  Skokov, V.A.: A remark on minimization methods that use the operation of space dilation. Kibernetika (Kiev), no. 4, 115–117 (1974)

97*  Slobodnik, S.G.: The continuity and differential properties of functions. Dis. D. Phil. Moscow 1966

98  Sonnevend, G.: On optimization of algorithms for function minimization. Colloq. Math. Soc. Janos Bolyai, 865–893 (1974)

99*  Tihonov, A.N., Glasko, V.B.: An application of the method of regularizations to nonlinear problems. Ž. Vycisl. Mat. i Mat. Fiz. 5, 463–473 (1975)

100*  Tikhonov, A.N., Samarskii, A.A.: Equations of mathematical physics. Gostekhizdat: Moscow 1951

101  Wolfe, P.: A method of conjugate subgradients for minimizing nondifferentiable functions. In: Math. Programming Stud. 3 (M.L. Balinski, P. Wolfe, eds.), pp. 145–173. North Holland: Amsterdam 1975

102*  Yudin, D.B.: Mathematical methods of control under incomplete information. Sovetskoe Radio: Moscow 1974

103*  Yudin, D.B., Nemirovskii, A.S.: Informational complexity and efficient methods for solving convex extremal problems. Ekonom. i Mat. Metody 12, 357–369 (1976)

104*  Zhurbenko, N.G.: Analysis of a class of algorithms for minimizing nonsmooth functions and their application to the solution of large scale problems. Diss. Doctor Philos. Kiev 1977

105*  Zhurbenko, N.G., Pinaev, E.G., Shor, N.Z., Yun, G.H.: The choice of the set of public service aircrafts and their distribution among airlines. Kibernetika (Kiev), no. 4, 138–141 (1976)

106  Zoutendijk, G.: Methods of feasible directions. Elsevier: Amsterdam 1960

# Additional References [2]

1  Aubin, J.P.: Mathematical methods of game and economic theory. North-Holland: Amsterdam 1979

2  Auslender, A.: Programmation convexe avec erreurs: méthodes d'epsilon-sous-gradients. Compt. Rend. Acad. Sci., Sér. A 284, 109–112 (1977)

3  Auslender, A.: Minimisation de fonctions localement lipschitziennes: applications à la programmation mi-convexe, mi-différentiable. In: Nonlinear Programming 3 (O.L. Mangasarian, R.R. Meyer, S.M. Robinson, eds.), pp. 429–460. Academic Press: New York 1978

4  Auslender, A.: Differential stability in nonconvex and nondifferentiable programming. In: Point-to-Set Maps in Mathematical Programming (P. Huard, ed.). Math. Programming Study 10, pp. 29–41. North-Holland: Amsterdam 1979

5  Bandler, J.W., Charalambous, C.: Nonlinear programming using minimax techniques. J. Optim. Theory Appl. 13, 607–619 (1974)

---

[2]  Not referred to within this volume.

6    Bazaraa, M.S., Shetty, C.M.: Nonlinear programming. Theory and applications. Wiley: New York 1979

7    Bertsekas, D.P.: Approximation procedures based on the method of multipliers. J. Optim. Theory Appl. 23, 487–510 (1977)

8    Charalambous, C.: Nonlinear least $p$-th optimization and nonlinear programming. Math. Programming 12, 195–225 (1977)

9    Chatelon, J., Hearn, D., Lowe, T.J.: A subgradient algorithm for certain minimax and minimum problems. Math. Programming 14, 130–145 (1978)

10   Cheney, E.W., Goldstein, A.A.: Newton's method for convex programming and Chebyshev approximation. Numer. Math. 1, 253–268 (1959)

11   Clarke, F.M.: Optimization and nonsmooth analysis. Wiley: New York 1983

12   Conn, A.R.: Constrained optimization using a nondifferentiable penalty function. SIAM J. Numer. Anal. 10, 760–784 (1973)

13   Danskin, J.M.: The theory of max-min. Springer: New York 1967

14   Dantzig, G.B.: Linear programming and extensions. Princeton University Press: Princeton 1963

15*  Demyanov, V.F.: Subgradient method and saddle points. Vest. Leningr. Univer. 13, 17–23 (1981)

16   Demyanov, V.F., Vasiliev, L.V.: Nondifferentiable optimization. Optimization Software Inc./Springer: New York (to appear, 1985). Russian edition: Nauka, Moscow (1981)

17   Dixon, L.C.W.: Reflections on nondifferentiable optimization, part I: the ball-gradient. J. Optim. Theory Appl. 32, 123–134 (1980)

18   Dixon, L.C.W., Gaviano, M.: Reflections on nondifferentiable optimization, part II: convergence. J. Optim. Theory Appl. 32, 259–276 (1980)

19   Eaves, B.C., Zangwill, W.I.: Generalized cutting plane algorithms. SIAM J. Control 9, 529–542 (1971)

20   Ekeland, I., Temam, R.: Analyse convexe et problèmes variationnels. Dunod: Paris 1974

21   Gauvin, J.: Shadow prices in nonconvex mathematical programming. Math. Programming 19, 300–312, 1980

22   Geoffrion, A.M.: Primal resource-directive approaches for optimizing nonlinear decomposable systems. Oper. Res. 18, 375–403 (1970)

23   Goffin, J.L.: On convergence rates of subgradient optimization methods. Math. Programming 13, 329–347 (1977)

24   Goffin, J.L.: Convergence rates of the ellipsoid method on general convex functions. Math. Oper. Res. 8, 135–150 (1983)

25   Goldstein, A.A.: Optimization of Lipschitz continuous functions. Math. Programming 13, 14–22 (1977)

26*  Gupal, A.M.: Stochastic methods for solving nonsmooth extremum problems. Kiev: Naukova Dumka 1979

27   Gwinner, J.: Bibliography on nondifferentiable optimization and non-smooth analysis. J. Comput. Appl. Math. 7, 277–285 (1981)

28   Hald, J., Madsen, K.: Combined LP and quasi-Newton methods for minimax optimization. Math. Programming 20, 49–62 (1981)

29   Han, S.-P.: Variable metric methods for minimizing a class of nondifferentiable functions. Math. Programming 20, 1–13 (1981)

30   Hiriart-Urruty, J.-B.: Generalized gradients of marginal value functions. SIAM J. Control Optim. 16, 301–316 (1978)

31   Hogan, W.W.: Directional derivatives of convex functions with applications to the completely convex case. Oper. Res. 21, 188–209 (1973)

32   Ioffe, A.D.: Nonsmooth analysis: differential calculus of nondifferentiable mappings. Trans. Amer. Math. Soc. 266, 1–56 (1981)

33   Kiwiel, K.C.: A phase I-phase II method for inequality constrained minimax problems. Control Cyb. 12, 55–75 (1983)

34    Kiwiel, K.C.: An aggregate subgradient method for nonsmooth convex minimiza-
      tion. Math. Programming 27, 320–341 (1983)
35    Lasdon, L.S.: Optimization theory for large systems. Macmillan: London 1970
36    Laurent, P.J.: Approximation et optimization. Herman: Paris 1972
37    Lemaréchal, C.: Bundle methods in nonsmooth optimization. In: Nonsmooth
      Optimization (C. Lemaréchal. R. Mifflin, eds.), pp. 79–102. Pergamon Press:
      Oxford 1978
38    Lemaréchal, C.: Nonsmooth optimization and descent methods. RR-78-4, Inter-
      national Institute for Applied Systems Analysis, Laxenburg, Austria 1978
39    Lemaréchal, C.: A view of line-searches. In: Optimization and Optimal Control (A.
      Auslender, W. Oettli, J. Stoer, eds.), pp. 59–78. Lecture Notes in Control and
      Information Sciences 30, Springer: Berlin 1981
40    Lemaréchal, C.: Numerical experiments in nonsmooth optimization. In: Progress
      in Nondifferentiable Optimization (E.A. Nurminski, ed.), pp. 61–84. CP-82-S8,
      International Institute for Applied Systems Analysis, Laxenburg, Austria 1982
41    Lemaréchal, C., Mifflin, R.: Global and superlinear convergence of an algorithm
      for one-dimensional minimization of convex functions. Math. Programming 24,
      241–256 (1982)
42    Lemaréchal, C., Strodiot, J.-J., Bihain, A.: On a bundle algorithm for nonsmooth
      minimization. In: Nonlinear Programming 4 (O.L. Mangasarian, R.R. Meyer, S.M.
      Robinson, eds.), pp. 245–281. Academic Press: New York 1981
43    Madsen, K.: An algorithm for minimax solution of over-determined systems of
      nonlinear equations. J. Inst. Math. Appl. 16, 321–328 (1975)
44    Madsen, K., Schjaer-Jacobsen, H.: Linearly constrained minimax optimization.
      Math. Programming 14, 208–223 (1978)
45    Marsten, R.E., Hogan, W.W., Blankenship, J.W.: The boxstep method for large-
      scale optimization. Oper. Res. 23, 389–405 (1975)
46    Mifflin, R.: A superlinearly convergent algorithm for one-dimensional constrained
      minimization with convex functions. Math. Oper. Res. 8, 185–195 (1983)
47    Mifflin, R.: Stationarity and superlinear convergence of an algorithm for univariate
      locally Lipschitz constrained minimization. Math. Programming 28, 50–71 (1984)
48*   Nemirovski, A.S., Yudin, D.B.: Complexity and efficiency of optimization methods.
      Nauka: Moscow 1979
49*   Nurminski, E.A.: Numerical methods for solving deterministic and stochastic
      minimax problems. Naukova Dumka: Kiev 1979
50    Nurminski, E.A., ed. progress in nondifferentiable optimization. CP-82-S8, Inter-
      national Institute for Applied Systems Analysis, Laxenburg, Austria 1982
51    Papavassilopoulos, G.: Algorithms for a class of nondifferentiable problems. J.
      Optim. Theory Appl. 34, 41–82 (1981)
52    Polak, E., Mayne, D.Q., Wardi, Y.: On the extension of constrained optimization
      algorithms from differentiable to nondifferentiable problems. SIAM J. Control
      Optim. 21, 179–203 (1983)
53    Pomerol, J.C.: The Lagrange multiplier set and the generalized gradient set of the
      marginal function of a differentiable program in a Banach space. J. Optim. Theory
      Appl. 38, 307–317 (1982)
54    Robinson, S.M.: A subgradient algorithm for solving K-convex inequalities. In:
      Optimization and Operations Research (P. Wolfe, ed.), pp. 237–245. Lecture Notes
      in Economics and Mathematical Systems 117, Springer: Berlin 1976
55    Rockafellar, R.T.: Lagrange multipliers and subderivatives of optimal value
      functions in mathematical programming. In: Nondifferential and Variational
      Techniques in Optimization (D.C. Sorensen, R.J.-B. Wets, eds.), pp. 28–66.
      Mathematical Programming Study 17, North-Holland: Amsterdam 1982
56    Strodiot, J.-J., Nguyen, V.H., Heukemes, N.: $\varepsilon$-Optimal solutions in nondifferen-
      tiable convex programming and some related questions. Math. Programming 25,
      307–328 (1983)

57 Topkis, D.M.: Cutting-plane methods without nested constraint sets. Oper. Res. 18, 404–413 (1970)
58 Topkis, D.M.: A cutting-plane algorithm with linear and geometric rates of convergence. J. Optim. Theory Appl. 36, 1–22 (1982)
59 Wierzbicki, A.P.: Lagrangian functions and nondifferentiable optimization. In: Progress in Nondifferentiable Optimization (E.A. Nurminski, ed.), pp. 173–213. CP-82-S8, International Institute for Applied Systems Analysis, Laxenburg, Austria 1982

# Subject Index

# QUADPACK

**A Subroutine Package for Automatic Integration**

By **R. Piessens, E. de Doncker-Kapenga, C. W. Überhuber, D. K. Kahaner**

1983. 26 figures. VIII, 301 pages. (Springer Series in Computational Mathematics, Volume 1).
ISBN 3-540-12553-1

**Contents:**

**Introduction**
**Theoretical Background:** Automatic Integration with QUADPACK. – Integration Methods.
**Algorithm Descriptions:** QUADPACK contents. – Prototype of Algorithm Description. – Algorithm Schemes. – Heuristics Used in the Algorithms.
**Guidelines for the Use of QUADPACK:** General Remarks. – Decision Tree for Finite-range Integration. – Decision Tree for Infinite-range Integration. – Numerical Examples. – Sample Programs Illustrating the Use of the QUADPACK Integrators.
**Special Applications of QUADPACK:** Two-dimensional Integration. – Hankel Transform. – Numerical Inversion of the Laplace Transform.
**Implementation Notes and Routine Listings:** Implementation Notes. – Routine Listings.
**References.**

QUADPACK presents a program package for automatic integration covering a wide variety of problems and various degrees of difficulty.

After a theoretical explanation of the quadrature methods, the algorithms used by the integrators are described, providing a detailed outline of the automatic integration strategies. The results for a set of parameter studies reveal efficiency and adequacy for wide ranges of problems. Applications are discussed for solving more complex problems, including double integration, computation of the Hankel transform, and inversion of the Laplace transform.

Apart from the explanation of the theory, the book includes the routine listings, the user's manual, and many detailed numerical examples and sample programs.

The documentation for use of the package is readable and clear for novice users. With the presentation of the mathematical methods and algorithms, however, some background in the area is assumed.

Springer-Verlag
Berlin
Heidelberg
New York
Tokyo

# Solving Elliptic Problems Using ELLPACK

By **J. Rice, R. F. Boisvert**

1984. 53 figures. Approx. 350 pages. (Springer Series in
Computational Mathematics, Volume 2).
ISBN 3-540-90910-9

**Contents:**

**The ELLPACK System:** Introduction. – The ELLPACK
Language. – Examples. – Advanced Language Facilities.
– Extending ELLPACK to Non-Standard Problems.
**The ELLPACK Modules:** The ELLPACK Problem Solv-
ing Modules. – ITPACK Solution Modules.
**The Performance of ELLPACK Software:** Performance
and its Evaluation. – The Model Problems. – Perfor-
mance of Modules to Discretize Elliptic Problems. –
Performance of Modules to Solve the Algebraic Equa-
tions.
**Contributor's Guide:** Software Parts for Elliptic Prob-
lems. – Interface Specifications. – Module Interface
Access. – Programming Standards. – Preprocessor Data.
**System Programmer's Guide:** Installing the ELLPACK
System. – Tailoring the ELLPACK System.
**Appendices:** The PDE Population. – The PG System. –
The Template Processor.

This book is a complete guide to the ELLPACK software
system solving elliptic partial differential equations.

ELLPACK consists of a very high level user interface to
over 50 problem solving modules. These modules are
state of the art software for two and three dimensional
problems and include finite difference, finite element,
SFT, multigrid and many other capabilities.

The book gives the practicing scientists the tools to solve
a wide range of elliptic problems with minimum effort. It
shows system programmers how to install and modify
ELLPACK and experts how to adapt ELLPACK to a
wide range of applications.

Springer-Verlag
Berlin
Heidelberg
New York
Tokyo

If you have any concerns about our products,
you can contact us on
ProductSafety@springernature.com

In case Publisher is established outside the EU,
the EU authorized representative is:
Springer Nature Customer Service Center GmbH
Europaplatz 3, 69115 Heidelberg, Germany

Printed by Libri Plureos GmbH
in Hamburg, Germany